K. Langanke   J. A. Maruhn   S. E. Koonin (Eds.)

Computational Nuclear Physics 1

Nuclear Structure

With 15 Figures, 19 Tables, and a Floppy Disk

Springer-Verlag

Berlin Heidelberg New York
London Paris Tokyo
Hong Kong Barcelona Budapest

Prof. K. Langanke
Institut für Theoretische Physik I
Universität Münster
Wilhelm-Klemm-Strasse 9, W-4400 Münster, FRG

Prof. J. A. Maruhn
Institut für Theoretische Physik
Universität Frankfurt
Robert-Mayer-Strasse 8–10, W-6000 Frankfurt 1, FRG

Prof. S. E. Koonin
Kellogg Radiation Laboratory
California Institut of Technology
Caltech
Pasadena, CA 91125, USA

The series *Computational Physics* was proposed and initiated
by K. Langanke.

ISBN 3-540-53571-3 Springer-Verlag Berlin Heidelberg New York

ISBN 0-387-53571-3 Springer-Verlag New York Berlin Heidelberg

56/3140-543210 - Printed on acid-free paper

Preface

Computation is essential to our modern understanding of nuclear systems. Although simple analytical models might guide our intuition, the complexity of the nuclear many-body problem and the ever-increasing precision of experimental results require large-scale numerical studies for a quantitative understanding.

Despite their importance, many nuclear physics computations remain something of a black art. A practicing nuclear physicist might be familiar with one or another type of computation, but there is no way to systematically acquire broad experience. Although computational methods and results are often presented in the literature, it is often difficult to obtain the working codes. More often than not, particular numerical expertise resides in one or a few individuals, who must be contacted informally to generate results; this option becomes unavailable when these individuals leave the field. And while the teaching of modern nuclear physics can benefit enormously from realistic computer simulations, there has been no source for much of the important material.

The present volume, the first of two, is an experiment aimed at addressing some of these problems. We have asked recognized experts in various aspects of computational nuclear physics to codify their expertise in individual chapters. Each chapter takes the form of a brief description of the relevant physics (with appropriate references to the literature), followed by a discussion of the numerical methods used and their embodiment in a FORTRAN code. The chapters also contain sample input and test runs, as well as suggestions for further exploration.

In choosing topics and authors, we have attempted to span the entire field, balancing modern implementations of venerable calculations on the one hand against speculative work at the cutting edge of present research on the other. Thus, in this first volume, concerned primarily with nuclear structure, we have included chapters on the shell model and RPA, as well as on boson and relativistic models. Similarly, the second volume, which deals primarily with nuclear reactions, will contain contributions on Hauser–Feshbach and DWBA calculations, as well as quark models of hadronic interactions. The level of presentation is generally that appropriate for a beginning graduate student who has a working knowledge of both elementary quantum mechanics and computational physics and might have taken (or be taking) a course in nuclear physics.

All of the codes discussed in this book are included on the disk. They have been written to conform to the FORTRAN 77 standard. In some cases, the authors have used subroutines commonly available in libraries (e.g., NAG or IMSL). When this has been done, we have been careful to fully describe the input, output, and purpose of these subroutines, so that the reader can substitute equivalents that might be more readily available. The computations are generally of a size and length that can be completed on a microcomputer like the IBM PS/2 model 70 in a tolerable amount of time, although a workstation or VAX-class machine is often preferable.

We have edited each chapter only very lightly, preferring instead to exhibit the pleasing diversity of the individual styles. However, we have ensured that each of the chapters is reasonably self-contained and that the codes do run as advertized. There are surely significant opportunities for the synergistic use of the various codes, but we have not explored these in detail.

We suspect that there are at least several uses for the material in this volume. At the simplest level, the codes can be run as black-boxes to produce pedagogical illustrations of various nuclear phenomena (for example, the shell-model oscillations in groundstate charge densities or the l-transfer dependence of DWBA cross sections). At a somewhat deeper level, the codes can be used as templates to explore the accuracy and efficiency of various numerical methods within the context of "real" problems. This collection of codes will also undoubtedly find broad use in the research community, and might even give our experimental colleagues some appreciation for the theorist's art.

We would like to express our appreciation for the enthusiasm and support that our colleagues have shown for this project, and offer thanks to Prof. Martin Zirnbauer for his help during the early phases of our work.

Pasadena   K. Langanke
October 1990   J. A. Maruhn
                                                                S. E. Koonin

Contents

List of Contributors

Åberg, Sven
Department of Mathematical Physics, Lund Institute of Technology
S-220 07 Lund, Sweden

Bengtsson, Tord
Department of Mathematical Physics, Lund Institute of Technology
S-220 07 Lund, Sweden

Bertsch, George
Department of Physics and Cyclotron Laboratory, Michigan State University
East Lansing, Michigan 48824, USA

Blok, Henk P.
Department of Physics and Astronomy, Vrije Universiteit, de Boelelaan
1081 HV Amsterdam, The Netherlands
and
NIKHEF-K, Postbus 4395
1009 AJ Amsterdam, The Netherlands

Carlson, Joseph A.
Los Alamos National Laboratory
Los Alamos, New Mexico 87545, USA

Glöckle, Walter
Institut für Theoretische Physik II, Ruhr-Universität Bochum
D-4630 Bochum, Federal Republic of Germany

Heisenberg, Jochen H.
Department of Physics, University of New Hampshire
Durham, New Hampshire 03824, USA

Hess, P.O.
(Fellow of the Deutscher Akademischer Austauschdienst)
Instituto de Ciencias Nucleares, Universidad Autónoma de México
México, D. F., México

Horowitz, C. J.
Physics Department and Nuclear Theory Center, Indiana University
Bloomington, Indiana 47405

Maruhn, Joachim A.
Institut für Theoretische Physik, Johann Wolfgang Goethe Universität
D-6000 Frankfurt, Federal Republic of Germany

Murdock, D.P.
Physics Department and Nuclear Theory Center, Indiana University
Bloomington, Indiana 47405
(Permanent address: Department of Physics, Box 5051, Tennessee
Technological University, Cookeville, TN 38505)

Ragnarsson, Ingemar
Department of Mathematical Physics, Lund Institute of Technology
S-220 07 Lund, Sweden

Reinhard, Paul-Gerhard
Universität Erlangen-Nürnberg
D-8520 Erlangen, Germany

Scholten, Olaf
Kernfysisch Versneller Instituut, Zernikelaan 25
9747 AA Groningen, The Netherlands

Serot, Brian D.
Physics Department and Nuclear Theory Center, Indiana University
Bloomington, Indiana 47405

Troltenier, Dirk
Institut für Theoretische Physik, Johann Wolfgang Goethe Universität
D-6000 Frankfurt, Federal Republic of Germany

Vallières, Michel
Department of Physics and Atmospheric Science, Drexel University
Philadelphia, PA 19104, USA

Wiringa, Robert B.
Physics Division, Argonne National Laboratory
Argonne, Illinois, USA

Wu, Hua
Department of Physics and Atmospheric Science, Drexel University
Philadelphia, PA 19104, USA
(Permanent address: Physics Department, McMaster University, Hamilton
Ontario, Canada L8S 4M1)

1. The Nuclear Shell Model

M. Vallières and H. Wu

1.1 Introduction

The shell-model approach played a fundamental role in our early under-standing of atoms and nuclei; in nuclear physics in particular, it has pro-vided us with our best understanding of the sd shell nuclei. To this day, it remains a fundamental approach in nuclear, atomic, and nonrelativistic quark physics, and it is considered a fundamental theoretical starting point in the derivation of most models applying to larger nuclei. As a numerical scheme, it has achieved considerable successes in both nuclear and atomic physics; however, it remains a notoriously difficult approach to implement when the number of particles and/or the number of shells or their specifying quantum numbers become large. In nuclear physics, it becomes difficult to apply for nuclei beyond the sd shell. For these nuclei we may well require a rethinking of the shell-model philosophy in which the collective subspace will be the aim of the calculations. The FDUO code, which we will describe below, is representative of this new trend in shell-model studies. It is de-signed to study nuclear collective behavior in a pair-truncated fermion space and yet it is implemented in a full shell-model-style algorithm. Furthermore it has already been successfully applied to medium and heavy nuclei to fit data. An added virtue for the purpose of this chapter is that the FDUO code is relatively short compared with the usual shell-model codes.

In Sect. 1.2 of this chapter we will first describe some of the shell-model codes currently used in nuclear physics; we will briefly indicate the methodology used in these codes as well as point out some advantages and disadvantages of each. We will then describe in some detail the FDUO code in Sect. 1.3. Section 1.4 will give a brief tutorial in the use of this code and provide specific examples.

1.2 Shell-Model Algorithms

The shell model describes each particle in a many-body system by a separate single-particle wave function. The potential which determines this wave function is that generated by the average motion of all of the other particles, and so depends on their single-particle wave functions. Normally only a subset of these particles is treated as active in one or a few major valence shells. These are then subjected to the influence of a one-body field and interact via an effective interaction. The shell-model codes must then solve the remaining many body-problem in this truncated space.

The different shell-model approaches that are used in nuclear physics can be summarized as indicated in Fig. 1.1. Of these schemes, the jj-coupling and the m-scheme can be viewed as the time-honored traditional schemes. The DUSM code on the other hand implements the newest shell-model algorithm; in this approach extensive use of the permutation group is made to build antisymmetric fermionic wave functions. The FDUO code, which will be described in detail in the following sections, represents an attempt based on the use of fermion-paired wave functions to describe numerically much larger nuclei than it was conventionally thought possible.

1.2.1   The jj-Coupling Scheme

This was the very first shell-model scheme proposed to describe nuclei [1.1]. Excellent pedagogical references describing the details of this approach can be found, in particular those by DeShalit and Talmi [1.2] and Brussard and Glaudemans [1.3]). The Oak Ridge shell-model code [1.4] was the first numerical implementation of this scheme in nuclear physics; it was recently recoded by Zwarts [1.5].

In this approach, the antisymmetric N-particle states in each shell are first built recursively, with $N_i \leq N$; then the multishell Hilbert space is constructed from these single-shell states. The steps involved can be listed very simply:

- Given a complete set of $N - 1$-particle antisymmetrized states in a given shell, build an (*overcomplete* and *not fully antisymmetrized*) set of N-particle angular-momentum-coupled states

$$\left[|j^{N-1}\alpha' J'\rangle|j\rangle\right]^J \tag{1.1}$$

- Use the coefficients of fractional parentage (CFPs)
$$\left[j^{N-1}\alpha' J' jJ|\}j^N\alpha J\right]$$
to form antisymmetrized N-particle states

$$|j^N\alpha J\rangle = \sum \left[j^{N-1}\alpha' J' jJ|\}j^N\alpha J\right]\left[|j^{N-1}\alpha' J'\rangle|j\rangle\right]^J . \tag{1.2}$$

- Repeat recursively for all particle numbers $N_i \leq N$ in each shell.
- Form multishell antisymmetric states by the outer product of antisymmetric states in each shell.
- Compute the expectation value of the Hamiltonian (1- and 2-body interaction terms) by insertion of complete sets of states for $N-1$ and $N-2$ systems and using the identity

$$\langle j^N\alpha J||a^\dagger||j^{N-1}\alpha' J'\rangle = \sqrt{N(2j+1)}\left[j^{N-1}\alpha' J' jJ|\}j^N\alpha J\right] . \tag{1.3}$$

- Diagonalize the Hamiltonian matrix (using the Lanczos algorithm [1.6] (see below) for large matrices).
- Compute the expectation values of the transition operators.

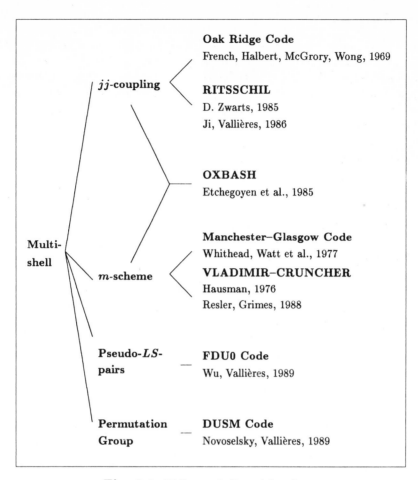

Fig. 1.1. Different shell-model codes.

The most time-consuming part of this algorithm is the calculation of the single-particle single-shell CFPs. The traditional algorithm for computing these CFPs dates back to Racah in 1949 [1.7], who explicitly antisymmetrized the last-particle wave function, and Redmond [1.8], who found recurrence relations which solve the Racah equations. However, these yield an *over-complete nonorthogonal* set of basis states. A Schmidt orthogonalization procedure (used in the original RITSSCHIL code [1.5]) can lead to numerical instabilities when the dimensions of the space become large. Ji and Vallières [1.9] proposed to diagonalize the Grammian matrix $G = \langle j^N \alpha' J | j^N \alpha J \rangle$ (the matrix of overlaps) whose large eigenvalues indicate states to be kept with norm equal to the eigenvalues and small eigenvalues indicate superfluous states to be discarded. But even this scheme has severe restrictions when the shell dimensions are large.

1.2.2   The m-scheme

Concurrently with the development of the jj-coupling codes, Whitehead et al. [1.10], the Lawrence Livermore group [1.11], and recently an Ohio State University group [1.12] developed the m-scheme approach. This algorithm exploits the similarity between second quantization and the binary operations on computers; that is a computer word is associated with each N-particle Slater determinant of the basis, using 1 and 0 to represent occupied and unoccupied orbits. Therefore the calculations of any matrix elements of second quantized operators involve only extremely efficient logical operations and the consultation of tables.

The obvious advantage of this scheme is that none of the complicated angular-momentum-coupling algebra needs to be performed explicitly. Note that the symmetries of the Hamiltonian are nevertheless preserved, i.e., states of good J will be generated by a rotationally invariant Hamiltonian, but that this will not be obvious until the matrix elements of relevant operators are taken, i.e., J^2 for rotationally invariant Hamiltonians.

On the other hand, the obvious disadvantage of this approach is that the dimensions get to be *extremely* large with increasing particle number. Another size limitation comes from the maximum number of orbitals imposed by the word length; multiple words can of course be combined but this results in far less efficient coding. It is also practically impossible to implement a truncation scheme of any type, short of diagonalizing the Casimir operators of some relevant group in this enormous space.

1.2.3   The Oxford–Buenos Aires Shell Model (OXBASH) Code

The OXBASH [1.13] code follows a hybrid algorithm between the m-scheme and the jj-scheme; it first builds the basis states in the m-scheme and then computes the Hamiltonian matrix in a jj-coupled scheme. This approach was devised to avoid the difficulties inherent in each angular-momentum scheme, namely to avoid angular-momentum-coupling algebra in setting up the jj-coupled basis and to avoid the very large matrices of the m-scheme. Of course, one has to pay the price for this apparent ease of computation, i.e. the difficulty in the transition from the m-scheme to the jj-coupled scheme.

1.2.4   The DUSM Code

The Drexel University Shell Model (DUSM) [1.14–17] code implements a new approach to perform shell-model calculations based on permutation-group concepts. It results from a collaboration of A. Novoselsky, R. Gilmore, J.Q. Chen, J. Katriel, and one of the present authors (M.V.). In this approach multiparticle multishell wave functions are built in each angular-momentum sector (i.e., in j and T (isospin space), L and S, or L, S, color, and flavor for quarks) and then combined globally to form antisymmetric wave functions; note that wave functions of arbitrary permutational symmetry are now required in each subspace. This coupling order is opposite to the coupling order used in the traditional approaches. The algorithm takes advantage of the fact that typical calculations are multishell in only one of the angular-momentum subspaces, i.e., nuclear calculations are multishell in j but single shell in isospin, and atomic calculations are multishell in l but single shell in spin, resulting in a very efficient scheme.

The calculation of the CFPs is performed by an efficient matrix diagonalization technique in which the quadratic Casimir operator of the subspace-dictated unitary group is diagonalized in the angular-momentum-coupled basis of the $N-1$ symmetrized states and the added last particle [1.14]. A generalization of the same algorithm is used to compute all of the necessary coupling coefficients between subspaces and between shells [1.15,16]. The matrices to be diagonalized are always relatively small since permutational symmetries can couple in very few ways to each other in a recursive build-up of many particle states. Furthermore, the construction and diagonalization of these matrices are easily vectorizable, making this an efficient algorithm for supercomputers.

The matrix elements of one- and two-body operators are easily computed once it is realized that the matrix elements of a creation and/or annihilation operator in any given shell are proportional to CFPs for any permutational symmetries of the parent and daughter states. The notion of path in Young-diagram space is introduced to help with the bookkeeping involved in these calculations; this helps greatly in sorting out the Young-diagrams in the various subspaces. Matrix elements of one- and two-body operators are then obtained by insertion of appropriate complete set of states of the $N-1$- and $N-2$-systems; this leads to the "summation over paths" technique [1.16].

The DUSM code is currently under development; all of the necessary coupling coefficient codes are written and fully tested, while the multishell code is near completion.

1.2.5   The Lanczos Algorithm

The Lanczos algorithm [1.6] is a method for diagonalizing large matrices when only a few extreme (low or high) eigenvalues and eigenstates are needed. It is an algorithm commonly used in all shell-model codes when the dimensions of the matrices become large. It has strictly nothing to do with the m-scheme, even though the matrices in that scheme are almost always large enough to require the use of the Lanczos algorithm; this is a common misconception.

The principle of this algorithm is as follows:

- Start with ϕ_1, an arbitrary Hilbert space vector containing components in the full space; often a *normalized* vector $\frac{1}{\sqrt{N_{\text{Hilbert}}}}(111...)$ is used.
- Build $H\phi_1 = \alpha_1\phi_1 + \beta_1\phi_2$.
- Continue building $H\phi_2 = \beta_1\phi_1 + \alpha_2\phi_2 + \beta_2\phi_3$.
- Repeat the previous step $\mathcal{N}$ times to build an $\mathcal{N} \times \mathcal{N}$ tridiagonal matrix (because of the Hermiticity of the Hamiltonian) $\begin{bmatrix} \alpha_1 & \beta_1 & & \\ \beta_1 & \alpha_2 & \beta_2 & \\ & \beta_2 & \alpha_3 & \beta_3 \\ & & \ddots & \ddots \end{bmatrix}$.
- Diagonalizing this matrix for $\mathcal{N} \ll N_{\text{Hilbert}}$ should provide a good approximation to the lowest eigenvalues of the original matrix.
- The eigenvectors are then reconstructed using the eigenvectors of the small matrix and the basis states.

1.2.6   Results in the 0s, 0p and sd Shells

The greatest successes of the nuclear shell model are undoubtedly in the sd shell. We refer here to nuclei from ^{16}O to ^{40}Ca for which the active or valence nucleons occupy the spherical sd major shell. The dimensions of the Hilbert space are not too large so that shell-model calculations are possible on computers of moderate size using the traditional algorithms. In particular, Wildenthal [1.18] has performed systematic shell-model studies of all the sd-shell nuclei in order to establish the best possible solution in this model space. He fitted the energies of all the known states of a given parity for all the nuclei in the shell, the parameters being the single-particle energies and two-body matrix elements, with a mild systematic mass dependence. The results of these calculations are well documented in review articles, in particular Ref. [1.18]. Suffice it to say that the calculations reproduce well all the systematic trends in the energy levels except for some deviations at the beginning or at the end of the major shell where we expect effects from adjacent shells. The transition rates and moments were also calculated.

The smaller p-shell nuclei can obviously be treated by the shell model. An early systematic set of calculations was performed by Cohen and Kurath [1.19] using a simple interaction. No-core calculations were performed in the m-scheme; in particular, Jastrow functions were used with the bare Reid potential to treat these nuclei in a combination of variational calculations with the shell-model approach [1.20]. One also has to be careful about center-of-mass spurious motion in these small nuclei [1.20].

1.3   The FDU0 Shell-Model Code

1.3.1   Beyond the sd Shell

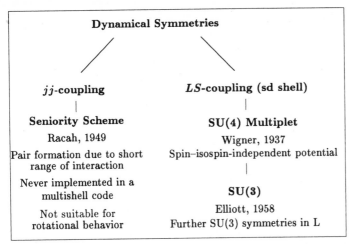

Fig. 1.2. Truncation schemes in the sd shell.

Beyond the sd shell lies a much more problematic region. Systematic studies of all states in these nuclei, similar to those in the sd shell, are no longer possible since the dimensions of the Hilbert spaces become prohibitive. Shell-model calculations beyond the sd shell require a rethinking of the shell-model philosophy; it is likely that only the collective states should become the focus of the calculations [1.21]. The tasks of the shell-model theorist therefore become the following:

- Select a proper truncation of the model space; it has to be drastic, to render the calculation possible, and systematic, to allow meaningful results.
- Derive an effective interaction to be used in the smaller model space.
- Develop efficient algorithms to perform these shell-model calculations, since the dimensions remain very large.

The first truncation of the model space is that of choosing an adequate set of valence orbitals based on the well-known existence of energy gaps between major shells; this is necessary but clearly not sufficient beyond the sd shell. This is when dynamical symmetries must be invoked in dictating further truncations compatible with the description of collective degrees of freedom. In this context, the only known such truncation scheme in jj-coupling is the seniority scheme, which applies well to vibrational nuclei. In order to describe rotational nuclei, models that contain an SU(3) group chain must be invoked. For nuclei in the sd shell, such models were proposed by Wigner [1.22], and then by Elliott [1.23], as indicated in Fig. 1.2.

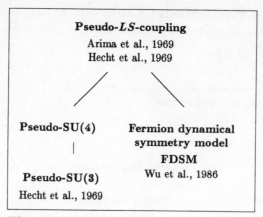

Fig. 1.3. Truncation schemes beyond the sd shell.

For nuclei beyond the sd shell, Hecht [1.24], and independently Arima et al. [1.25], introduced the idea of an LS-coupling via the idea of pseudo-angular-momenta and spins. Two schemes implement this idea, the pseudo-SU(3)-scheme which is a direct generalization of Elliott's SU(3) scheme for the sd shell, and the Fermion dynamical symmetry model (FDSM) [1.26]; these are illustrated in Fig. 1.3. The latter approach models the collective subspace using fermion pairs of angular momentum zero and two. FDUO is a shell-model code written by the present authors [1.27,28] which implements the truncation of the shell-model space as dictated by the FDSM.

1.3.2 The Background to FDSM

Following Ginocchio [1.29] the spherical single-particle orbitals are mapped onto a pseudo-angular-momentum, k, and pseudo-spin, i, by a unitary transformation, $j = k + i$, yielding

$$b^\dagger_{km_k im_i} = \sum C^{jm_j}_{km_k im_i} a^\dagger_{jm_j}. \tag{1.4}$$

Pairs of fermions are then formed in which the pseudo-angular-momenta (spins) are coupled to angular momentum 0 or 2, the complementary spins (angular momenta) being coupled to zero; i.e.

$$A^{r\dagger}_m = \sum_{ki} \sqrt{\Omega_{ki}/2} \left(b^\dagger_{ki} b^\dagger_{ki} \right)^{r0}_{m0} \tag{1.5}$$

or

$$A^{r\dagger}_m = \sum_{ki} \sqrt{\Omega_{ki}/2} \left(b^\dagger_{ki} b^\dagger_{ki} \right)^{0r}_{0m}, \tag{1.6}$$

where $r=0,2$ and $\Omega_{ki} = \frac{1}{2}(2k+1)(2i+1)$. The k and i values can only be 1 (the k active scheme) or 3/2 (i active) if only S and D pairs are allowed [1.29]. The FDSM [1.26] further requires that the LS unitary transformation, i.e., $j=k+i$, maps the entire fermion space to the k–i basis when applied to one physical major shell under the condition that only one type of symmetry be used; this restricts the choice of k and i to unique values in each shell.

Besides the pair operators, multipole operators are also constructed:

$$p_u^r = \sum_i \sqrt{\Omega_{ki}/2} \left(b_{1i}^\dagger \tilde{b}_{1i}\right)_{\mu 0}^{r0} \tag{1.7}$$

and

$$p_u^r = \sum_k \sqrt{\Omega_{ki}/2} \left(b_{k3/2}^\dagger \tilde{b}_{k3/2}\right)_{0\mu}^{0r}. \tag{1.8}$$

In practice, the sum over k can be dropped since a unique value is generally sufficient to map the fermion space to the i-active scheme. In each scheme, the pair creation and annihilation operators and multipole operators form closed algebras: Sp(6) is generated if k is active and SO(8) if i is active. These groups have very little in common; their group-chain decompositions for instance are very different. However, their canonical commutation relations differ in a relatively minor way; i.e.

$$\left[A_\mu^r, A_\nu^{s\dagger}\right] = \Omega \delta_{rs} \delta_{\mu\nu} - 2 \sum_t K_{r-\mu,s\nu}^{t\sigma}(-)^\mu P_\sigma^t , \tag{1.9}$$

$$\left[P_\mu^r, A_\nu^{s\dagger}\right] = \sum_t K_{r\mu,s\nu}^{t\sigma} A_\sigma^{t\dagger} , \tag{1.10}$$

$$\left[P_\mu^r, P_\nu^s\right] = \frac{1}{2} \sum_t [(-)^t - (-)^{r+s}] K_{r\mu,s\nu}^{t\sigma} P_\sigma^t , \tag{1.11}$$

where $\Omega = \frac{1}{2}\sum(2j+1)$, with the sum performed over the normal parity levels of the major shell, and

$$K_{r\mu,s\nu}^{t\sigma} = K_{r,s}^t C_{r\mu s\nu}^{t\sigma} , \tag{1.12}$$

$$K_{r,s}^t = (-)^\alpha \sqrt{(2\alpha+1)(2r+1)(2s+1)} \left\{ \begin{array}{ccc} r & s & t \\ \alpha & \alpha & \alpha \end{array} \right\} , \tag{1.13}$$

with $\alpha = 1$ for k active and $\alpha = \frac{3}{2}$ for i active. This suggests a possible numerical implementation of this scheme in shell-model language, where the multiparticle multishell wave functions are built out of pairs rather than single fermions. The approach then becomes a *full shell-model technique performed in a systematically truncated fermion space* according to the philosophy of the FDSM. The truncation of the fermion pair space to only pairs of angular momentum 0 and 2 is referred to as the $u=0$ subspace. The FDU0 program [1.27,28] implements this truncation scheme.

1.3.3   The Basis

The building of the basis in a pair-coupled scheme parallels that of the jj-coupling scheme for single fermions and is best expressed in a flow chart as shown in Fig. 1.4. An *overcomplete nonorthogonal* basis for $N + 1$-pair states is obtained by coupling the angular momentum of the last pair of fermions to the angular momentum of the N-pair wave functions,

$$|N + 1, rbLJM_J\rangle = \sum_{M_L, m_r} C^{JM_J}_{LM_L r m_r} A^{r\dagger}_{m_r} |NbLM_L\rangle \ . \tag{1.14}$$

The commutation relations are used to express the $N + 1$-overlaps in terms of the N-overlaps and multipole operators' matrix elements; this guarantees the proper antisymmetrization of the wave functions,

$$
\begin{aligned}
G &\equiv \langle N + 1, r'b'L'JM_J | N + 1, rbLJM_J\rangle \\
&= \Omega_1 \delta_{rr'} \delta_{LL'} \delta_{bb'} \\
&\quad + \sum_{cs} (-1)^{s+J-L-L'} U \begin{bmatrix} r' & s & L \\ r & J & L' \end{bmatrix} \frac{1}{\sqrt{2L'+1}} \frac{1}{\sqrt{2L+1}} \\
&\quad \times \langle Nb'L' \| A^{r\dagger} \| N - 1, cs\rangle \langle NbL \| A^{r'\dagger} \| N - 1, cs\rangle \\
&\quad - 2 \sum_{t} K^t_{r'r} U \begin{bmatrix} L & t & L' \\ r & J & r \end{bmatrix} \\
&\quad \times (-1)^{r'} \sqrt{\frac{2t+1}{2r+1}} \frac{1}{\sqrt{2L'+1}} \langle Nb'L' \| P^t \| NbL\rangle \ .
\end{aligned}
\tag{1.15}
$$

The diagonalization of the Grammian matrix [1.9] allows the selection and recursive build-up of an orthogonal basis,

$$
\begin{aligned}
|N + 1, aJM_J\rangle &= \sum_{bLr} W^{aJ}_{bLr} |N + 1, rbLJM_J\rangle \\
&= \sum_{bLr} \frac{Z^{aJ}_{bLr}}{\sqrt{e_{aJ}}} |N + 1, rLJM_J\rangle \ ,
\end{aligned}
\tag{1.16}
$$

where Z^{aJ}_{bLr} is an eigenvector of the overlap matrix corresponding to the counting label a and e_{aJ} is its nonzero eigenvalue; (bLr) specify the N-pair states and the last pair operators' quantum numbers and W^{aJ}_{bLr} are the expansion coefficients.

In order to make the algorithm recursive we must also be able to compute the matrix elements of the pair and multipole operators. We get for the pair operators

$$
\begin{aligned}
&\frac{1}{\sqrt{2J+1}} \langle N + 1, aJ \| A^{r\dagger} \| NbL\rangle \\
&= \langle N + 1, aJM_J | N + 1, rbLJM_J\rangle \\
&= \sum_{b'L'r'} W^{aJ}_{b'L'r'} \langle N + 1, r'b'L'JM_J | N + 1, rbLJM_J\rangle
\end{aligned}
\tag{1.17}
$$

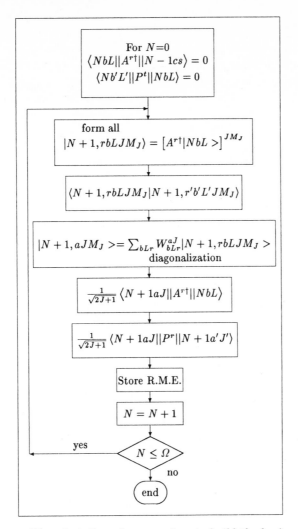

Fig. 1.4. Iterative procedure to build the basis.

and for the multipole operators

$$
\frac{1}{\sqrt{2J'+1}}\langle N+1,a'J'\|P^r\|N+1,aJ\rangle
$$

$$
= \sum_{bLIb'L'} W_{bLI}^{aJ} U \begin{bmatrix} r & L & L' \\ I & J' & J \end{bmatrix} (-1)^{L-L'+J-J'}
$$

$$
\times \frac{1}{\sqrt{2J'+1}}\langle N+1,a'J\|A^{I\dagger}\|NbL'\rangle \frac{1}{\sqrt{2L'+1}}\langle Nb'L'\|P^r\|NbL\rangle
$$

$$
+ \sum_{bLIu} W_{bLI}^{aJ} U \begin{bmatrix} L & I & J \\ r & J' & u \end{bmatrix} (-1)^{r+I-u} K_{rI}^u
$$

$$\times \frac{1}{\sqrt{2J'+1}} \langle N+1, a'J' \| A^{u\dagger} \| NbL \rangle \ . \tag{1.18}$$

The details of the derivation of these formulae can be found in Ref. [1.28]. These are sufficient to start the iterative scheme, provided we use (N=0)

$$\langle NbL \| A^{r\dagger} \| N-1, cs \rangle = 0 \tag{1.19}$$

and

$$\langle NbL \| P^t \| NbL \rangle = 0 \ . \tag{1.20}$$

1.3.4 The Hamiltonian Matrix

A very general form for the Hamiltonian is obtained by allowing any pairing and multipole interactions to take place within both the neutron and the proton sectors, and allowing a general multipole interaction to couple the two sectors:

$$H - E_0 = \sum_{\lambda=\pi,\nu} \left[\sum_{r=0,2} G_\lambda^r A^{r\dagger}(\lambda) A^r(\lambda) + \sum_{r=1}^{2\alpha(\lambda)} B_\lambda^r P^r(\lambda) P^r(\lambda) \right]$$
$$+ \sum_{r=1}^{2\alpha_0} B_{\pi\nu}^r P^r(\pi) P^r(\nu) \ , \tag{1.21}$$

where $\alpha_0 = \min(\alpha(\pi), \alpha(\nu))$. The first and second terms on the r.h.s. are the pairing and multipole interactions for identical particles while the third term is the neutron–proton interaction.

The identical-particle-interaction terms can all be calculated beforehand in the multipair basis and stored; in this part of the Hamiltonian matrix, elements then become simple linear combinations of elementary matrix elements:

$$\langle Na'JM_J | A^{r\dagger} A^r | NaJM_J \rangle = \sum_{bJ'} \frac{1}{\sqrt{2J+1}} \frac{1}{\sqrt{2J+1}}$$
$$\times \langle Na'J \| A^{r\dagger} \| N-1, bJ' \rangle \langle NaJ \| A^{r\dagger} \| N-1, bJ' \rangle; \tag{1.22}$$

and

$$\langle Na'JM_J | P^r P^r | NaJM_J \rangle$$
$$= \sum_{bJ'} \frac{1}{\sqrt{2J+1}} \langle Na'J \| P^r \| NbJ' \rangle \frac{1}{\sqrt{2J+1}} \langle NaJ \| P^r \| NbJ' \rangle \ . \tag{1.23}$$

The total neutron–proton system wave functions are expanded in the SO(3) coupled neutron and proton basis:

$$|N_\nu a_\nu J_\nu; N_\pi a_\pi J_\pi; JM_J \rangle$$
$$= \sum_{M_\nu M_\pi} C_{J_\nu M_\nu J_\pi M_\pi}^{JM_J} |N_\nu a_\nu J_\nu M_\nu \rangle |N_\pi a_\pi J_\pi M_\pi \rangle \ . \tag{1.24}$$

In this basis, the matrix elements for the Hamiltonian are

$$\langle N_\nu a'_\nu J'_\nu; N_\pi a'_\pi J'_\pi; JM_J | H | N_\nu a_\nu J_\nu; N_\pi a_\pi J_\pi; JM_J \rangle = \sum_{\lambda=\pi,\nu} \delta_{a'_\sigma a_\sigma} \delta_{J'_\sigma J_\sigma} \delta_{J'_\lambda J_\lambda}$$

$$\langle N_\lambda a'_\lambda J_\lambda M_\lambda | \left[\sum_{r=0,2} G^r_\lambda A^{r\dagger}(\lambda) A^r(\lambda) + \sum_{r=1}^{2\alpha(\lambda)} B^r_\lambda P^r(\lambda) P^r(\lambda) \right] | N_\lambda a_\lambda J_\lambda M_\lambda \rangle$$

$$+ \sum_{r=1}^{2\alpha_0} B^r_{\pi\nu} U \begin{bmatrix} J_\nu & r & J'_\nu \\ J'_\pi & J & J_\pi \end{bmatrix} \sqrt{\frac{2J'_\pi + 1}{2J_\pi + 1}} (-1)^r \frac{1}{\sqrt{2J'_\nu + 1}} \frac{1}{\sqrt{2J'_\pi + 1}}$$

$$\langle N_\nu a'_\nu J'_\nu \| P^r(\nu) \| N_\nu a_\nu J_\nu \rangle \langle N_\pi a'_\pi J'_\pi \| P^r(\pi) \| N_\pi a_\pi J_\pi \rangle \qquad ,$$

$$(1.25)$$

where σ is the complement of λ, that is $\sigma = \pi$, if $\lambda = \nu$ and $\sigma = \nu$ if $\lambda = \pi$.

Once the Hamiltonian matrix is constructed, the energy spectra and wave functions are obtained by a straightforward diagonalization procedure, either via a standard IMSL routine (see Sect. 1.5) or via the Lanczos algorithm [1.6] depending on the dimension of the subspace for a given angular momentum.

1.3.5 Matrix Elements for Transition Operators

The $M1$ and $E2$ effective transition operators are defined as

$$M1 = g_\pi J_\pi + g_\nu J_\nu \ , \tag{1.26}$$
$$E2 = e_\pi P^2(\pi) + e_\nu P^2(\nu) \ , \tag{1.27}$$

where J_π and J_ν are the angular-momentum operators ($J = \sqrt{4\alpha} \times \sqrt{(\alpha+1)/3}\, P^1$) and $P^2(\pi)$ and $P^2(\nu)$ are the quadrupole operators in the proton and neutron sectors, respectively; g_π, g_ν, e_π, and e_ν are the proton and neutron pair's gyromagnetic ratios and effective charges. The $B(E2)$ transition rates and quadrupole moments, Q, are then

$$B(E2 : |i\rangle \rightarrow |f\rangle) = \frac{2J_f + 1}{2J_i + 1} S^2 \ , \tag{1.28}$$

$$Q = \sqrt{\frac{16\pi}{5}} C^{JJ}_{JJ20} S \ . \tag{1.29}$$

The $|i\rangle$ and $|f\rangle$ stand for the initial and the final states, J_i and J_f being the corresponding angular momenta. J is the angular momentum for the state for which the quadrupole moment is to be computed ($J_i = J_f = J$); S is defined as

$$S = e_\pi \frac{1}{\sqrt{2J_f + 1}} \langle f \| P^2(\pi) \| i \rangle + e_\nu \frac{1}{\sqrt{2J_f + 1}} \langle f \| P^2(\nu) \| i \rangle \ , \tag{1.30}$$

where

$$\frac{1}{\sqrt{2J_f+1}}\langle f\|P^2(\pi)\|i\rangle$$

$$= \sum_{a_\nu J_\nu a'_\pi J'_\pi a_\pi J_\pi} Z_f(a_\nu J_\nu a'_\pi J'_\pi) Z_i(a_\nu J_\nu a_\pi J_\pi)\, U \begin{bmatrix} J_\nu & J_\pi & J_i \\ 2 & J_f & J'_\pi \end{bmatrix}$$

$$\times \frac{1}{\sqrt{2J'_\pi+1}}\langle N_\pi a'_\pi J'_\pi\|P^2(\pi)\|N_\pi a_\pi J_\pi\rangle \tag{1.31}$$

and

$$\frac{1}{\sqrt{2J_f+1}}\langle f\|P^2(\nu)\|i\rangle$$

$$= \sum_{a_\nu J_\nu a'_\nu J'_\nu a_\pi J_\pi} Z_f(a'_\nu J'_\nu a_\pi J_\pi) Z_i(a_\nu J_\nu a_\pi J_\pi)\, U \begin{bmatrix} J_\pi & J_\nu & J_i \\ 2 & J_f & J'_\nu \end{bmatrix}$$

$$\times (-1)^{J_\nu - J'_\nu + J_f - J_i} \frac{1}{\sqrt{2J'_\nu+1}}\langle N_\nu a'_\nu J'_\nu\|P^2(\nu)\|N_\nu a_\nu J_\nu\rangle \ . \tag{1.32}$$

The factors Z_i and Z_f are the initial- and final-state wave-function amplitudes.

Similar formulae hold for the $M1$ transitions and magnetic dipole moments, μ:

$$B(M1:|i\rangle \to |f\rangle) = \frac{2J_f+1}{2J_i+1} S^2 \ , \tag{1.33}$$

$$\mu = \sqrt{\frac{4\pi}{3}} C^{JJ}_{JJ10} S \ , \tag{1.34}$$

$$S = g_\pi \frac{1}{\sqrt{2J_f+1}}\langle f\|\boldsymbol{J}_\pi\|i\rangle + g_\nu \frac{1}{\sqrt{2J_f+1}}\langle f\|\boldsymbol{J}_\nu\|i\rangle \ , \tag{1.35}$$

$$\frac{1}{\sqrt{2J_f+1}}\langle f\|\boldsymbol{J}_\pi\|i\rangle = \sum_{a_\nu J_\nu a_\pi J_\pi} Z_f(a_\nu J_\nu a_\pi J_\pi) Z_i(a_\nu J_\nu a_\pi J_\pi)$$

$$\times U \begin{bmatrix} J_\nu & J_\pi & J_i \\ 1 & J_f & J_\pi \end{bmatrix} \sqrt{J_\pi(J_\pi+1)} \ , \tag{1.36}$$

$$\frac{1}{\sqrt{2J_f+1}}\langle f\|\boldsymbol{J}_\nu\|i\rangle = \sum_{a_\nu J_\nu a_\pi J_\pi} Z_f(a_\nu J_\nu a_\pi J_\pi) Z_i(a_\nu J_\nu a_\pi J_\pi)$$

$$\times U \begin{bmatrix} J_\pi & J_\nu & J_i \\ 1 & J_f & J_\nu \end{bmatrix} (-1)^{J_f - J_i}\sqrt{J_\nu(J_\nu+1)} \ . \tag{1.37}$$

1.3.6 The Fitting Procedure

The FDUO code is capable of fitting energy spectra. Mathematically, one minimizes the weighted least-squares function

$$\phi \equiv \chi^2 = \sum_i w_i (E_i^{th} - E_i^{exp})^2 \, , \tag{1.38}$$

where i sums over all energy levels of interest, w_i are the weight factors for each level, E_i^{th} are the theoretical predictions and E_i^{exp} are the experimental energy levels. The simplest way to minimize ϕ is to regard it as a general nonlinear function of the parameters x_α of the model and invoke the usual minimization procedure, generally found in any software library. However, this could be very time consuming; for example, the gradient method requires the evaluation of ϕ at least $f + 1$ times, with f the number of adjustable parameters, for every iteration leading to a new set of parameters.

For many years, there has existed a very efficient fitting procedure in the implementation of shell-model calculations. In this procedure, one first guesses a good starting point for the parameters, say x_α^0, then finds the corresponding wave functions, $|i\rangle^0$, and then predicts a new set of parameters using

$$E_i^{th} = {}^0\langle 0|H(x_\alpha)|0\rangle^0 \, . \tag{1.39}$$

This is an exact result when $x_\alpha = x_\alpha^0$; for $x_\alpha \neq x_\alpha^0$, this is only valid up to first-order perturbation. This leads to a very efficient fitting procedure, especially when H is a linear function of the adjustable parameters (as is the case for the shell model).

The previous fitting procedure is valid when the wave functions change slowly as the parameters change. Given this restriction, it is perhaps the most efficient way to fit the given energies and is the first choice in the FDUO program package. However, we found that in some cases this procedure fails to converge owing to rapid changes of the wave functions. When this happens, the code automatically switches to another procedure in which the derivatives are evaluated analytically:

$$\frac{\partial \phi}{\partial x_\alpha} = 2 \sum_i w_i (E_i^{th} - E_i^{exp}) \frac{\partial E_i^{th}}{\partial x_\alpha} \tag{1.40}$$

with

$$\frac{\partial E_i^{th}}{\partial x_\alpha} = \frac{\partial \langle i|H|i\rangle}{\partial x_\alpha} = \left\langle i \left| \frac{\partial H}{\partial x_\alpha} \right| i \right\rangle \tag{1.41}$$

yielding

$$\frac{\partial \phi}{\partial x_\alpha} = 2 \sum_i w_i (E_i^{th} - E_i^{exp}) \left\langle i \left| \frac{\partial H}{\partial x_\alpha} \right| i \right\rangle \, . \tag{1.42}$$

The gradient of ϕ at any point x_α in the f-dimensional parameter space is consequently evaluated with a single calculation of the wave functions. If the Hamiltonian is linear in x_α, $H = \sum_\alpha x_\alpha H_\alpha$, then

$$\frac{\partial \phi}{\partial x_\alpha} = 2 \sum_i w_i (E_i^{\text{th}} - E_i^{\text{exp}}) \langle i | H_\alpha | i \rangle \ . \tag{1.43}$$

Once the gradient of ϕ is found, the new parameter values are predicted by

$$x_\alpha := x_\alpha - \frac{\partial \phi}{\partial x_\alpha} \mathrm{d}x / |\nabla \phi| \ , \tag{1.44}$$

where $\mathrm{d}x$ is the distance between the old and new parameter points in the parameter space. This fitting procedure predicts new parameters for every solution of the Schrödinger equation and is f-times more efficient than procedures which compute the derivatives numerically.

1.3.7 An FDU0 Primer

Installation

The FDU0 package contains four major codes in separate FORTRAN source files: FDSMCFP.FOR, PD.FOR, FDU0.FOR, and FDTR.FOR, and a library file called LIB.FOR. In addition the diagonalization and linear-equation-solver routines are called from the IMSL single- and double-precision libraries (see Sect. 1.5). The executable codes are built as follows:

1. Compile all five FORTRAN source files separately.

2. Link the FDSMCFP object file to the LIB object file and the *double-precision* IMSL object library.

3. Link the PD, FDU0, and FDTR object files separately to the LIB object file and the *single-precision* IMSL object library.

Once the executable files are produced, run FDSMCFP first. This code prompts you for a symmetry type; respond $SO8$ for the first time. When finished, restart FDSMCFP and give $SP6$ as input; this will take longer to finish but you will need to run it only once. Now run the program PD twice for the $SO8$ and $SP6$ symmetries respectively. These runs will produce several data files (CFPs and matrix elements) required by the programs FDU0 and FDTR.

Spectrum Computation

The program FDU0 diagonalizes the FDSM Hamiltonian

$$\begin{aligned} H \ = \ & EN * J_\nu (J_\nu + 1) + B2N * P_\nu^2 \, P_\nu^2 + B3N * P_\nu^3 \, P_\nu^3 \\ & + G0N * S_\nu^\dagger \, S_\nu + G2N * D_\nu^\dagger \, D_\nu \\ & + EP * J_\pi (J_\pi + 1) + B2P * P_\pi^2 \, P_\pi^2 + B3P * P_\pi^3 \, P_\pi^3 \\ & + G0P * S_\pi^\dagger \, S_\pi + G2P * D_\pi^\dagger \, D_\pi \\ & + ENP * \boldsymbol{J}_\nu \cdot \boldsymbol{J}_\pi + B2NP * P_\nu^2 \, P_\pi^2 + B3NP * P_\nu^3 \, P_\pi^3 \\ & + B1N * P_\nu^1 \, P_\nu^1 + B1P * P_\pi^1 \, P_\pi^1 + B1NP * P_\nu^1 \, P_\pi^1 \\ & + EJ * J(J + 1) \end{aligned} \tag{1.45}$$

in the $u = 0$ space. Here JN, JP, and J are the neutron, proton, and total angular momenta respectively; $P_\nu^1, P_\nu^2, P_\nu^3$ and $P_\pi^1, P_\pi^2, P_\pi^3$ are the multipole operators; $S_\nu^\dagger, S_\nu, S_\pi^\dagger, S_\pi$ and $D_\nu^\dagger, D_\nu, D_\pi^\dagger, D_\pi$ are the pair operators. The model space is determined by the shell pair degeneracies OMP and OMN, the symmetry type $NEUTRON\_SYMMETRY$ and $PROTON\_SYMMETRY$, and the normal pair numbers $N1P$ and $N1N$.

The input parameters are given in an input file `FDSM.INP` with only one namelist block *inp*. Parameters are classified in four groups, with the following allowed and default values:

Model Space Parameters:

name	allowed values	default
NEUTRON_SYMMETRY	'SO8','SP6'	'SP6'
PROTON_SYMMETRY	'SO8','SP6'	'SO8'
OMN,OMP	2,6,10 for SO8 symmetry	10
	6,15,21 for SP6 symmetry	15
N1N	0 to OMN	0
N1P	0 to OMP	0

Hamiltonian Parameters:

name	allowed values	default
all parameters in (1.45)	real numbers	0.0

Spectrum Fitting Parameters:

name	allowed values	default
LEVELS(1:50)	real number with only one decimal digit	0.0
EE(1:50)	real numbers	0
WEIGHT(1:50)	real numbers	1.0

The array $LEVELS$ stores up to 50 levels to which the spectrum is being fitted, EE and $WEIGHT$ containing the corresponding energies and weighting factors. For example, to fit the third 4^+ state with energy 2.23 and weight 2.5, you would specify $LEVELS(2) = 4.3, EE(2) = 2.23, WEIGHT(2) = 2.5$.

Control Parameters:

name	allowed values	default
NJ	non-negative integer	8
MAXJ	non-negative integer	10
WAVE	'ON','OFF'	'OFF'
LAN	positive integer	50
FIT	non-negative integer	0
FIT_MODE	'G','L'	'L'
EPS	small positive real number	0.0
STEP	positive real number	automatically decided

where:

NJ is an output control for the number of levels to be printed for each J value.

MAXJ is the maximum J value for which eigenvalues or eigenstates are to be found.

WAVE is the wave-function production control; if 'ON', the file FDSM.WAV is produced.

LAN is the LANCZOS-diagonalization-procedure matrix size; the LANCZOS procedure will be invoked when NS, the dimension of the subspace for a given J, is greater than MAXDNS (default: 120).

FIT is the number of fitting iterations: for no fitting, set FIT=0; when FIT> 0, only the Hamiltonian parameters with initial nonzero values will be adjusted.

FIT_MODE controls the fitting method: 'L' for linear fitting, 'G' for gradient method. However, if the 'L' method fails to converge, the 'G' method will automatically take over.

EPS is a fitting exit-control parameter; if TD, the total derivative of the χ^2, is less than EPS, then the program stops.

STEP is the expected change of distance in the parameter space in one iteration in the 'G' fitting mode.

Example Run

The following is an example of FDSM.INP

```
$INP
NEUTRON_SYMMETRY='SP6'
PROTON_SYMMETRY='SP6'
OMN=15
OMP=15
N1N=3
N1P=2
WAVE='ON'
NJ=8
MAXJ=2
FIT=10
EPS=1.5
STEP=0.1
B2NP=-2.1,B2N=-1.1,B2P=-0.9,B1NP=-2.05,B1N=-1.2,B1P=-0.8
LEVELS=0.1,0.2,0.5,1.1,1.2,1.4,2.1,2.2,2.3
EE=0,27,42,15,36,40.5,0,15,27
$END
```

The output text file FDSM.OUT produced by FDUO using the above input file is:

```
****************INPUT IMAGE***********************
$INP
NEUTRON_SYMMETRY='SP6'
```

```
PROTON_SYMMETRY='SP6'
 OMN=15
 OMP=15
 N1N=3
 N1P=2
 WAVE='ON'
 NJ=8
 MAXJ=2
 FIT=10
 EPS=1.5
 STEP=0.1
 B2NP=-2.1,B2N=-1.1,B2P=-0.9,B1NP=-2.05,B1N=-1.2,B1P=-0.8
 LEVELS=0.1,0.2,0.5,1.1,1.2,1.4,2.1,2.2,2.3
 EE=0,27,42,15,36,40.5,0,15,27
 $END
 **************************************************
 J= 0
0.0000   27.0721   28.1969   29.1289   43.5459   44.0667   44.8737   52.2161
 J= 1
15.6702   37.6305   38.5070   41.1816   43.1704   49.7528   51.0743   54.4467
 J= 2
-0.0217   15.6419   26.8960   27.1860   28.1300   28.2435   28.9842   29.1992
 FIRST 0+ STATE ENERGY=  -67.600
 FI=   6.390909

  New set of parameters:
 $INP
 NEUTRON_SYMMETRY          = 'SP6',
 PROTON_SYMMETRY = 'SP6',
 OMN     =          15,
 OMP     =          15,
 N1P     =           2,
 N1N     =           3,
 EP      = -0.3776203    ,
 B2P     =  -1.000746    ,
 B3P     =  0.0000000E+00,
 G0P     =  0.0000000E+00,
 G2P     =  0.0000000E+00,
 EN      = -0.3751329    ,
 B2N     = -0.9998565    ,
 B3N     =  0.0000000E+00,
 G0N     =  0.0000000E+00,
 G2N     =  0.0000000E+00,
 ENP     = -0.7533147    ,
 B2NP    =  -1.999742    ,
 B3NP    =  0.0000000E+00,
 B1P     =  0.0000000E+00,
 B1N     =  0.0000000E+00,
 B1NP    =  0.0000000E+00,
 EJ      =  0.0000000E+00,
 WAVE    = 'ON ',
 NJ      =           8,
 LAN     =         100,
```

```
MAXJ     =              2,
MINJ     =              0,
FIT      =              9,
EPS      =    1.500000      ,
STEP     =  0.1000000       ,
FIT_MODE= 'L',
LEVELS  = 0.1000000, 0.2000000, 0.5000000,   1.100000, 1.200000,
1.400000, 2.100000,  2.200000,  2.300000, 41*0.0000000E+00,
EE       = 0.0000000E+00, 27.00000, 42.00000, 15.00000,
36.00000, 40.50000, 0.0000000E+00, 15.00000, 27.00000,
41*0.0000000E+00,
WEIGHT  = 50*1.000000
$END
J= 0
  0.0000 26.9950 27.0010 27.0037 41.9933 41.9963 42.0001 50.9884
J= 1
 14.9993 35.9941 36.0009 40.4962 40.4999 47.9952 48.0004 52.4922
J= 2
 -0.0077 14.9944 26.9846 26.9901 26.9914 26.9944 26.9969 26.9983
FIRST 0+ STATE ENERGY=  -64.999
FI=   4.4596949E-04
minimum reached with total derivative=   1.394188
```

In the output file, $FI = \chi^2 = \sum_i WEIGHT(i)[E^{\text{th}} - EE(i)]^2$. The level energies are given relative to the first 0^+, while the real first 0^+-state energy is shown separately.

Transition Rates and Moments

Upon setting WAVE='ON' in FDSM.INP, FDUO will record the information about the input parameters, the basis quantum numbers, and the wave functions in the file FDSM.WAV. The program FDTR computes the transitions; it requires an input file FDTR.INP. The first line of this file is for a comment, followed by the namelist block inp which can contain the following parameters:

1. All model-space parameters.

2. All Hamiltonian parameters printed out in FDSM.OUT.

3. Control KEY with allowed value 0 or 1 and default 1.

4. Effective charges EFFN, EFFP and g-factors GRN, RGP. The default values are 1.0.

If $KEY = 0$, the parameter sets (1) and (2) are ignored. Instead, the ones read from the file FDSM.WAV will be used. If $KEY = 1$, a matching check is made between the parameter sets (1), (2), and the file FDSM.WAV; the program will continue only if the data sets are identical.

Following the namelist block inp, put a text line

```
lm      from      to
```

followed by any number of transition-identifier lines. A transition identifier contains an integer number lm, allowed to be 1 ($M1$) or 2 ($BE2$), and two real numbers *from* and *to*, each with only one decimal digit representing the initial and final states for the transition in the form $J.n$ where J stands for the angular momentum and n for the occurrence ordering. For example, $from = 2.3$ means the third $J = 2$ state. A transition identifier

```
2     2.1     2.2
```

requests FDTR to compute the $BE(2)$ transition from the first 2^+ state to the second 2^+ state. In the case that $from = to$, the moment of that state will be calculated. The following is an example of a file FDTR.INP

```
$INP
KEY=0
EFFN=0.5,EFFP=1
$END
lm   from    to
2  0.1    2.1
2  2.1  2.2
1 2.1 2.1
```

The output is a text file named FDTR.OUT (using the wave functions produced in the last section):

```
Input parameters:
************************************************************************
NEUTRON_SYMMETRY=SP6      PROTON_SYMMETRY=SP6
OMN=15      OMP=15      N1N= 3      N1P= 2
EN= -0.3751  B2N= -0.9999  B3N=  0.0000  GON=  0.0000  G2N=  0.0000
EP= -0.3776  B2P= -1.0007  B3P=  0.0000  GOP=  0.0000  G2P=  0.0000
ENP= -0.7533   B2NP= -1.9997   B3NP=  0.0000
Q2=  0.5000*P2(N)+  1.0000*P2(P)      M1=  1.0000*J(N)+  1.0000*J(P)
************************************************************************
LM    FROM              TO          SN        SP         RESULT
2  0.1(  0.000)  2.1( -0.008)    2.1630    1.4426   31.8546(BE2)
2  2.1( -0.008)  2.2( 14.994)    0.3863   -0.3851    0.0368(BE2)
1  2.1( -0.008)  2.1( -0.008)    1.4675    0.9819    4.0933(dipole)
```

The file FDTR.OUT first mirrors the input parameters. Then the transition identifiers, the energy values for the related levels, and the results will be displayed line by line. Some intermediate results (SN,SP) are also printed, as they have very simple relations with the final results and are independent of $EFFN, EFFP, GRN$, and GRP.

$$S = \begin{cases} EFFN * SN + EFFP * SP & LM = 2 \\ GRN * SN + GRP * SP & LM = 1 \end{cases}$$

$$transition = \frac{2J_f + 1}{2J_i + 1} S^2$$

$$moment = \begin{cases} \sqrt{\frac{4\pi}{3}} C_{JJ10}^{JJ} S & LM = 1 \\ \sqrt{\frac{16\pi}{5}} C_{JJ20}^{JJ} S & LM = 2. \end{cases}$$

With these relations, one can obtain transition rates with the same wave functions but different effective $E2$ or $M1$ operators by running FDTR only once.

Implementation

FDUO was developed on VAX computers. The current version was also ported and tested on Convex and IBM-3090 mainframes. It is currently being ported to SUN and MIPS workstations.

Suggestions

The first use of this algorithm was to establish theoretical results: when the neutrons and protons are in shells which dictate different dynamical groups, the usual group decomposition techniques fail to provide a proof of the existence of rotational behavior. The code was then used to obtain such an approximate rotational band using a generic "quadrupole-quadrupole" Hamiltonian. The program was also used to obtain approximate vibrational spectra in neutron–proton systems of arbitrary group structure. These calculations are described in Ref. [1.27]; it would be very instructive for the reader to reproduce some of these.

The procedure can also be used to "fit" energy spectra and transition rates. The FDUO code is very efficient and can compute energies and transition rates for any even–even nucleus within the restricted FDSM model space [1.28].

1.4   Conclusions

The shell-model approach is a very exciting way to study nuclei; it is considered by most a very fundamental approach that has had great successes in describing small nuclei. We have described in this chapter the shell-model code FDUO which can be used for any large even–even nucleus in the periodic table. It is based on a specific truncation of the model space according to FDSM and yet reflects well the new trends in trying to apply shell-model techniques to larger nuclei. Note that the DUSM code aims at the same goal as FDUO, i.e., to describe large nuclei; however, DUSM is not locked into a specific pair structure and furthermore can compute even–odd and odd–odd nuclei as easily as the even–even ones. By simple changes to its input files DUSM can in fact perform calculations in any LS (or JT) schemes, offering the possibility of dynamically selecting the best model-space truncation and pair formation. The current phenomenological successes of FDUO and FDSM make it exciting to pursue the development of the more ambitious DUSM code.

1.5   Technical Note

As discussed in Sect. 1.3.7, the code calls some IMSL subroutines. In the
following we give a brief description of the IMSL subroutines most of which
are required in the program FDUO. The subroutine DEVCSF, which is called in
the program FDMSCFP, needs double precision. A more complete description
of the routines is to be found in the IMSL^{TM} MATH/LIBRARY User's
Guide (Edition 1.1).

EVCSB

Purpose:		To compute all of the eigenvalues and eigenvectors of a real symmetric matrix in band-symmetric storage mode
Usage:		CALL EVCSB (N,A,LDA,NCODA,EVAL,EVEC,LDEVEC)
Arguments:		
N	–	Order of matrix A (input)
A	–	Band-symmetric matrix (input)
LDA	–	Leading dimension of A exactly as specified in calling program (input)
NCODA	–	Number of codiagonals (input)
EVAL	–	Vector (length N) containing eigenvalues in increasing order (output)
EVEC	–	Matrix of order N containing eigenvectors (output); Jth eigenvector is in Jth column
LDEVEC	–	Leading dimension of EVEC exactly as specified in calling program (input)

EVLSB

Purpose:		To compute all of the eigenvalues of a real symmetric matrix in band-symmetric storage mode
Usage:		CALL EVLSB (N,A,LDA,NCODA,EVAL)
Arguments:		
N	–	Order of matrix A (input)
A	–	Band-symmetric matrix (input)
LDA	–	Leading dimension of A exactly as specified in calling program (input)
NCODA	–	Number of codiagonals (input)
EVAL	–	Vector (length N) containing eigenvalues in increasing order (output)

LSARG

Purpose:		To solve a real general system of linear equations with iterative refinements
Usage:		`CALL LSARG (N,A,LDA,B,IPATH,X)`
Arguments:		
N	–	Number of equations (input).
A	–	$N \times N$ matrix containing the coefficients of the linear system (input)
B	–	Vector of length N containing the right-hand side of the linear system (input)
IPATH	–	Path indicator (input) IPATH=1 means that the system AX=B is solved IPATH=2 means that the system trans(A)X=B is solved where trans( ) indicates the transpose
X	–	Vector (length N) containing the solution of the linear system (output)

E2CFS

Purpose:		To compute all of the eigenvalues and eigenvectors of a real symmetric matrix
Usage:		`CALL E2CFS (N,A,LDA,NCODA,EVAL,EVEC,LDEVEC,WORK)`
Arguments:		
N	–	Order of matrix A (input)
A	–	Real symmetric matrix (input)
LDA	–	Leading dimension of A exactly as specified in calling program (input)
EVAL	–	Vector (length N) containing eigenvalues in increasing order (output)
EVEC	–	Matrix of order N containing eigenvectors (output); Jth eigenvector is in Jth column
LDEVEC	–	Leading dimension of EVEC exactly as specified in calling program (input)
WORK	–	Work array of length N

E2LFS

Purpose:		To compute all of the eigenvalues of a real symmetric matrix
Usage:		CALL E2LSF (N,A,LDA,EVAL,ACOPY,WORK)
Arguments:		
N	–	Order of matrix A (input)
A	–	Real symmetric matrix (input)
LDA	–	Leading dimension of A exactly as specified in calling program (input)
EVAL	–	Vector (length N) containing eigenvalues in increasing order (output)
ACOPY	–	Work matrix of size $N \times N$
WORK	–	Work array of length $2 * N$

DEVCSF

Purpose:		To compute all of the eigenvalues and eigenvectors of a real symmetric matrix
Usage:		CALL DEVCSF (N,A,LDA,EVAL,EVEC,LDEVEC)
Arguments:		
N	–	Order of matrix A (input)
A	–	Real symmetric matrix (input)
LDA	–	Leading dimension of A exactly as specified in calling program (input)
EVAL	–	Vector (length N) containing eigenvalues in increasing order (output)
EVEC	–	Real matrix of size $N \times N$ (output) The Jth eigenvector, corresponding to EVAL(J), is stored in the Jth column
LDEVEC	–	Leading dimension of EVEC exactly as specified in calling program (input)

Acknowledgments

It is a pleasure to acknowledge our colleagues and collaborators: J.-Q. Chen, D.H. Feng, R. Gilmore, X. Ji, J. Katriel, and A. Novoselsky. We are particularly grateful to D.W.L. Sprung for having read this manuscript carefully. Supercomputer time from the National Center for Supercomputer Applications (NCSA) in Illinois is acknowledged. This work was partially supported by the National Science Foundation PHY88-43235 and by Drexel University Louis and Bessie Stein Family Foundation. Finally, one of us (M.V) is grateful for the hospitality of the Weizmann Institute for a short stay during which this paper was completed.

References

[1.1] M.G. Mayer, Phys. Rev. **75** (1949) 1968; O.J. Haxel, J.H.D. Jensen, and H.E. Suess, Phys. Rev. **75** (1949) 1766; M.G. Mayer and J.H.D. Jensen, *Elementary Theory of Nuclear Shell Structure*, (Wiley, New York, 1955)

[1.2] A. De-Shalit and I. Talmi, *Nuclear Shell Theory* (Academic, New York, 1963)

[1.3] P.J. Brussard and P.W.M. Glaudemans, *Shell Model Applications in Nuclear Spectroscopy* (North-Holland, Amsterdam, 1977)

[1.4] J.B. French, E.C. Halbert, J.B. McGrory, and S.S.M. Wong, Adv. Nuc. Phys. **3** (1969) 193; E.C. Halbert, J.B. McGrory, B.H. Wildenthal, and S.P Pandya, Adv. Nuc. Phys. **4** (1971) 375; see also J.B. McGrory, B.H. Wildenthal, Ann. Rev. Nucl. Part. Sci. **30** (1980) 383

[1.5] D. Zwarts, Comp. Phys. Comm. **38** (1985) 565

[1.6] C. Lanczos, J. Res. Nat. Bur. Stand. **45** (1950) 55

[1.7] G. Racah, Phys. Rev. **63** (1943) 367

[1.8] P.J. Redmond, Proc. Roy. Soc. **A222** (1954) 84

[1.9] X. Ji and M. Vallières, Phys. Rev. **C35** (1987) 1583

[1.10] R.R. Whitehead, A. Watt, B.J. Cole, and I. Morrison, Adv. Nuc. Phys. **9** (1977) 123

[1.11] (VLADIMIR code) R.F. Hausman, Jr. Lawrence Livermore National Laboratory Report No UCRL-52178, Ph. D. Thesis, 1976

[1.12] (CRUNCHER code) D.A. Resler and S.M. Grimes, Computers in Physics, May–June (1988) 65

[1.13] A. Etchegoyen, W. D. M. Rae, N. S. Godwin, W. A. Richter, C. H. Zimmerman, B. A. Brown, W. E. Ormand, and J. S. Winfield, MSU-NSCL report 524 (1985)

[1.14] A. Novoselsky, J. Katriel, and R. Gilmore, J. Math. Phys. **29** (1988) 1368

[1.15] A. Novoselsky, M. Vallières, and R. Gilmore, Matrix Elements of Shell Model Hamiltonians in Multiple Angular Momentum Coupling Schemes, Phys. Rev. **C38** (1988) 1440

[1.16] J.-Q. Chen, A. Novoselsky, M. Vallières, and R. Gilmore, A New Approach to Multi-shell Calculations in Multiple Angular Momentum Coupling Schemes, Phys. Rev. **C39** (1989) 1088

[1.17] M. Vallières, A. Novoselsky, and R. Gilmore, "New Shell-Model Algorithm" in *Proceedings of the International Conference on Computers in Physics*, edited by L. Deyuan and D.H. Feng, (World Scientific, Singapore, 1989)

[1.18] B.H. Wildenthal, *Progress in Particle and Nuclear Physics*, edited by D. Wilkinson (Pergamon, Oxford, 1984); B.H. Wildenthal in *Nuclear Shell Models*, edited by M. Vallières and B.H. Wildenthal (World Scientific, Singapore, 1984); B.A. Brown and B.H. Wildenthal, Ann. Rev. Nucl. Part. Sci. **38** (1988)

[1.19] S. Cohen and D. Kurath, Nucl. Phys. **73** (1965) 1; S. Cohen and D. Kurath, Nucl. Phys. **A141** (1970) 145; D. Kurath, Phys. Rev. **C7** (1973) 1390; D. Kurath and D.J. Millener, Nucl. Phys. **A238** (1975) 269

[1.20] J.M. Irvine, G.S. Mani, and M. Vallières, Czech. J. Phys. **B** (1974) 24; J.M. Irvine, G.S. Mani, V.F. Pucknell, M. Vallières, and F. Yazici, Ann. Phys. (N.Y.) **102** (1976) 129

[1.21] X. Ji, B.H. Wildenthal, and M. Vallières, Nucl. Phys. **A492** (1989) 815

[1.22] E.P. Wigner, Phys. Rev. **51** (1937) 106

[1.23] J.P. Elliott, Proc. Roy. Soc. **A245** (1958) 128

[1.24] K. Hecht and A. Adler, Nucl. Phys. **A137** (1969) 129

[1.25] A. Arima, M. Harvey, and K. Shimuzu, Phys. Lett., **B30** (1969) 517

[1.26] C.L. Wu, D.H. Feng, X.G. Chen, J.Q. Chen, and M.W. Guidry, Phys. Lett. **168B** (1986) 313; Phys. Rev. **C36** (1987) 1157

[1.27] Hua Wu and M. Vallières, Phys. Rev. **C39** (1989) 1066; FDU0 code, Hua Wu, unpublished

[1.28] Hua Wu, Ph. D. Thesis, Drexel University, 1989

[1.29] J.N. Ginocchio, Ann. Phys. (N.Y.) **126** (1980) 234

2. The Skyrme–Hartree–Fock Model of the Nuclear Ground State

P.-G. Reinhard

2.1 Introduction

Two decades ago, with the introduction of Skyrme forces [2.1], Hartree–Fock calculations became feasible in nuclear physics. Since then, they have been applied to a great variety of phenomena, including deformation properties, superheavy nuclei, vibrations, and heavy-ion collisions [2.2]. Nonetheless, their most straightforward application, the description of the ground state of spherical nuclei, remains a useful tool. It serves as the basis for many further applications in nuclear-structure physics; e.g., for studying refinements and variants of the force [2.3], for understanding electron-scattering data, for describing hyperons in nuclei, or for RPA vibrations of the ground-state. Thus, it is desirable to have a code optimized for speed. Such a code also provides a good example of the fast numerical techniques that are necessary for large scale applications.

The spherical Hartree–Fock code presented here now is more than 12 years old. It has been rewritten often to keep up with developing programming standards. We hope that it now represents a fairly modern FORTRAN style and that it is sufficiently commented. From the beginning the code was optimized for speed. For this purpose we have developed improved variants of the gradient iteration [2.4–2.7]. Finite-difference schemes with five-point precision are used for the wave functions. We find this the optimum choice in one-dimensional applications, although this decision may change for higher dimensions. The code has been applied intensively in least-squares fits for optimizing the Skyrme–force parameters [2.3], for calculating ground-state correlations [2.8], and for nuclear-excitation studies. Although we have omitted most of these branches, options, and extensions here, the present code is applicable to quite a body of nuclear-structure physics. We shall try to give a complete but short account of all necessary ingredients.

The chapter is outlined as follows. In Sect. 2.2, we describe Hartree–Fock theory with Skyrme–forces and the schematic treatment of pairing. In Sect. 2.3, we explain the numerical representation of wave functions and fields, and we discuss the shell ordering that defines the bookkeeping. In Sect. 2.4, we present the iteration schemes for solving the coupled equations. In Sect. 2.5, we explain the computation of the charge density and related quantities, and in Sect. 2.6 we give a short overview of the code structure, the input, and the installation.

2.2 Skyrme–Hartree–Fock Theory with Schematic Pairing

The Hartree–Fock equations and the pairing equations will be derived variationally from the total energy functional of the nucleus. Thus the whole model and all its variants can be presented by discussing the various contributions to the energy functional

$$E = E_{\text{Skyrme}} + E_{\text{Coulomb}} + E_{\text{pair}} - E_{\text{cm}} , \qquad (2.1)$$

where E_{Skyrme}, the energy functional of the Skyrme force, is the leading part. The Coulomb energy, E_{Coulomb}, seems straightforward, but some approximations will be needed to achieve a simplicity similar to E_{Skyrme}. A schematic pairing is defined by E_{pair}, and finally a correction for the spurious center-of-mass motion of the mean field is subtracted with E_{cm}. We discuss each of these contributions in the following subsections.

2.2.1 The Skyrme Energy Functional

The Skyrme Force

The Skyrme force is an effective force for nuclear Hartree–Fock calculations that aims to parametrize the t-matrix for nucleon–nucleon scattering in the nuclear medium in a simple and efficient manner. It is a zero-range, density- and momentum-dependent force of the form

$$
\begin{aligned}
V_{\text{Skyrme}} = \ & t_0(1 + x_0 P_x)\delta(\boldsymbol{r}_i - \boldsymbol{r}_j) \\
& + \tfrac{1}{2}t_1(1 + x_1 P_x)\{\boldsymbol{p}_{12}^2\delta(\boldsymbol{r}_i - \boldsymbol{r}_j) + \delta(\boldsymbol{r}_i - \boldsymbol{r}_j)\boldsymbol{p}_{12}^2\} \\
& + t_2(1 + x_2 P_x)\boldsymbol{p}_{12} \cdot \delta(\boldsymbol{r}_i - \boldsymbol{r}_j)\boldsymbol{p}_{12} \\
& + \tfrac{1}{6}t_3(1 + x_3 P_x)\rho^\alpha(\bar{\boldsymbol{r}})\delta(\boldsymbol{r}_i - \boldsymbol{r}_j) \\
& + \mathrm{i}t_4\boldsymbol{p}_{12} \cdot \delta(\boldsymbol{r}_i - \boldsymbol{r}_j)(\boldsymbol{\sigma}_i + \boldsymbol{\sigma}_j) \times \boldsymbol{p}_{12} ,
\end{aligned}
\qquad (2.2)
$$

where $\boldsymbol{p}_{12} = \boldsymbol{p}_i - \boldsymbol{p}_j$ is the relative momentum, P_x the space exchange operator $\boldsymbol{r}_i \leftrightarrow \boldsymbol{r}_j$, $\boldsymbol{\sigma}$ the vector of Pauli spin matrices, and $\bar{\boldsymbol{r}} = \tfrac{1}{2}(\boldsymbol{r}_i + \boldsymbol{r}_j)$. The simplicity of the ansatz allows the expectation value of the energy for Slater determinants to be evaluated in terms of a few densities and currents:

$$
\begin{aligned}
\rho_q(\boldsymbol{r}) &= \sum_{\beta \in q} w_\beta \, \varphi_\beta(\boldsymbol{r})^+ \varphi_\beta(\boldsymbol{r}) \\
\boldsymbol{j}_q(\boldsymbol{r}) &= \frac{\mathrm{i}}{2} \sum_{\beta \in q} w_\beta \left[\nabla\varphi_\beta(\boldsymbol{r})^+ \varphi_\beta(\boldsymbol{r}) - \varphi_\beta(\boldsymbol{r})^+ \nabla\varphi_\beta(\boldsymbol{r}) \right] \\
\tau_q(\boldsymbol{r}) &= \sum_{\beta \in q} w_\beta \, \nabla\varphi_\beta(\boldsymbol{r})^+ \cdot \nabla\varphi_\beta(\boldsymbol{r}), \\
\nabla\boldsymbol{J}_q(\boldsymbol{r}) &= -\mathrm{i} \sum_{\beta \in q} w_\beta \, \nabla\varphi_\beta(\boldsymbol{r})^+ \cdot \nabla \times \boldsymbol{\sigma}\varphi_\beta(\boldsymbol{r}) ,
\end{aligned}
\qquad (2.3)
$$

where φ_β is the single-particle wave function of state β, and the isospin label q runs over $q \in \{pr, ne\}$ (pr = proton and ne = neutron). The occupation probability of the state β is denoted by w_β. Completely filled shells have $w_\beta = 1$, but fractional occupancies occur for nonmagic nuclei; these are determined by the pairing scheme discussed in Sect. 2.2.3.

The Energy in Spherical Representation

In the following we restrict consideration to the (stationary) ground state of spherical nuclei. The single-particle wave functions can then be separated as

$$\varphi_\beta(\boldsymbol{r}) = \frac{R_\beta(r)}{r} \mathcal{Y}_{j_\beta l_\beta m_\beta}(\theta, \phi) \,. \tag{2.4}$$

The functions $\mathcal{Y}_{j_\beta l_\beta m_\beta}$ are spinor spherical harmonics [2.9]. The radial wave functions R_β are independent of the m_β quantum number. The factor r^{-1} has been seperated to simplify the overlap integrals to simple r-integration,

$$\langle \varphi_\beta | \varphi'_\beta \rangle = \delta_{j_\beta j'_\beta} \delta_{l_\beta l'_\beta} \delta_{m_\beta m'_\beta} \int_0^\infty dr R_\beta(r) R'_\beta(r) \,, \tag{2.5}$$

and to simplify the kinetic energy in the radial Schrödinger equation to a second derivative. All shells are assumed to be filled equally over the m_β-degeneracy, so that the densities become spherical. Altogether the resulting Skyrme energy functional becomes

$$
\begin{aligned}
E_{\text{Skyrme}} = \ & 4\pi \int_0^\infty dr\, r^2 \Bigg\{ \frac{\hbar^2}{2m}\tau + \tfrac{1}{2}t_0(1 + \tfrac{1}{2}x_0)\rho^2 - \tfrac{1}{2}t_0(\tfrac{1}{2} + x_0)\sum_q \rho_q^2 \\
& + \tfrac{1}{12}t_3(1 + \tfrac{1}{2}x_3)\rho^{\alpha+2} - \tfrac{1}{12}t_3(\tfrac{1}{2} + x_3)\rho^\alpha \sum_q \rho_q^2 \\
& + \tfrac{1}{4}[t_1(1 + \tfrac{1}{2}x_1) + t_2(1 + \tfrac{1}{2}x_2)]\rho\tau \\
& - \tfrac{1}{4}[t_1(\tfrac{1}{2} + x_1) - t_2(\tfrac{1}{2} + x_2)]\sum_q \rho_q\tau_q \\
& - \tfrac{1}{16}[3t_1(1 + \tfrac{1}{2}x_1) - t_2(1 + \tfrac{1}{2}x_2)]\rho\nabla^2\rho \\
& + \tfrac{1}{16}[3t_1(1 + \tfrac{1}{2}x_1) + t_2(1 + \tfrac{1}{2}x_2)]\sum_q \rho_q\nabla^2\rho_q \\
& - \tfrac{1}{2}t_4[\rho\nabla\boldsymbol{J} + \sum_q \rho_q\nabla\boldsymbol{J}_q] \Bigg\} \,,
\end{aligned} \tag{2.6}
$$

where $\nabla^2 = \partial_r^2 + \frac{2}{r}\partial_r$ and ∂_r is shorthand for $\frac{\partial}{\partial r}$. Note that the current $\boldsymbol{j}$ vanishes for stationary states. The densities in spherical representation are

$$\rho_q(r) = \sum_{n_\beta j_\beta l_\beta} w_\beta \frac{2j_\beta + 1}{4\pi} \left(\frac{R_\beta}{r} \right)^2 \,,$$

$$\tau_q(r) = \sum_{n_\beta j_\beta l_\beta} w_\beta \frac{2j_\beta + 1}{4\pi} \left[\left(\partial_r \frac{R_\beta}{r} \right)^2 + \frac{l(l+1)}{r^2} \left(\frac{R_\beta}{r} \right)^2 \right] , \tag{2.7}$$

$$\nabla J_q(r) = \left(\partial_r + \frac{2}{r} \right) J_q(r) ,$$

$$J_q(r) = \sum_{n_\beta j_\beta l_\beta} w_\beta \frac{2j_\beta + 1}{4\pi} [j_\beta(j_\beta + 1) - l_\beta(l_\beta + 1) - \tfrac{3}{4}] \frac{2}{r} \left(\frac{R_\beta}{r} \right)^2 ,$$

where the occupation weights w_β are also independent of m_β. The densities without an isospin label in (2.6) represent total densities, summed over both species:

$$\rho = \rho_{\text{pr}} + \rho_{\text{ne}} \quad , \quad \tau = \tau_{\text{pr}} + \tau_{\text{ne}} \quad , \quad \nabla J = \nabla J_{\text{pr}} + \nabla J_{\text{ne}}$$

The Hartree–Fock Equations

The Hartree–Fock equations for the radial wave functions R_β are obtained by varying the energy functional with respect to R_β under the constraint that the wave functions R_β be orthonormal. This yields the single-particle Schrödinger-like equation

$$h_q R_\beta = \epsilon_\beta R_\beta \tag{2.8}$$

with the mean-field Hamiltonian

$$h_q = \partial_r \mathcal{B}_q \partial_r + U_q + U_{ls,q} \boldsymbol{l} \cdot \boldsymbol{\sigma} , \tag{2.9}$$

where

$$\mathcal{B}_q = \frac{\hbar^2}{2m_q} + \tfrac{1}{8}[t_1(1 + \tfrac{1}{2}x_1) + t_2(1 + \tfrac{1}{2}x_2)]\rho$$
$$\qquad - \tfrac{1}{8}[t_1(\tfrac{1}{2} + x_1) - t_2(\tfrac{1}{2} + x_2)]\rho_q , \tag{2.10}$$

$$U_q = t_0(1 + \tfrac{1}{2}x_0)\rho - t_0(\tfrac{1}{2} + x_0)\rho_q$$
$$\qquad + \tfrac{1}{12}t_3\rho^\alpha \left[(2 + \alpha)(1 + \tfrac{1}{2}x_3)\rho - 2(\tfrac{1}{2} + x_3)\rho_q \right.$$
$$\qquad \left. - \alpha(\tfrac{1}{2} + x_3)\frac{\rho_{\text{pr}}^2 + \rho_{\text{ne}}^2}{\rho} \right]$$
$$\qquad + \tfrac{1}{4}[t_1(1 + \tfrac{1}{2}x_1) + t_2(1 + \tfrac{1}{2}x_2)]\tau$$
$$\qquad - \tfrac{1}{4}[t_1(\tfrac{1}{2} + x_1) - t_2(\tfrac{1}{2} + x_2)]\tau_q$$
$$\qquad - \tfrac{1}{8}[3t_1(1 + \tfrac{1}{2}x_1) - t_2(1 + \tfrac{1}{2}x_2)]\Delta\rho$$
$$\qquad + \tfrac{1}{8}[3t_1(\tfrac{1}{2} + x_1) + t_2(\tfrac{1}{2} + x_2)]\Delta\rho_q$$
$$\qquad - \tfrac{1}{2}t_4(\nabla J + \nabla J_q) + U_{\text{Coul}} , \tag{2.11}$$

$$U_{ls,q} = \tfrac{1}{4}t_4(\rho + \rho_q) + \tfrac{1}{8}(t_1 - t_2)J_q - \tfrac{1}{8}(x_1t_1 + x_2t_2)J , \tag{2.12}$$

and where the Coulomb contribution to the potential U_{Coul} will be given in the next subsection. The densities ρ, τ, and ∇J are given in (2.7). Note that (2.8) is nonlinear in the wave functions R_β via the density-dependent mean-field Hamiltonian (2.9–2.12).

The Rearrangement Energy

Once the solution to the Hartree–Fock equations (2.8–2.12) has been found (possibly with pairing for the w_β) one could evaluate the energy of the system straightforwardly from the functional (2.6). However, most of the energy information is already contained in the single-particle energies ϵ_β. Thus, it is simpler to compute the energy via

$$E_{\text{Skyrme}} + E_{\text{Coulomb}} = \frac{1}{2}\left(E_{\text{kin}} + \sum_\beta w_\beta \epsilon_\beta \right) + E_{\text{rearr}} \tag{2.13}$$

$$E_{\text{kin}} = 4\pi \sum_q \int_0^\infty dr r^2 \frac{\hbar}{2m_q}\tau_q \tag{2.14}$$

$$E_{\text{rearr}} = -4\pi \int_0^\infty dr\, r^2 \frac{\alpha}{24} t_3 \rho^\alpha \left[(1+\tfrac{1}{2}x_3)\rho^2 - (\tfrac{1}{2}+x_3)(\rho_{\text{pr}}^2 + \rho_{\text{ne}}^2) \right]$$
$$+ E_{\text{Coul,rearr}} . \tag{2.15}$$

The combination of "half single-particle energies and half kinetic energy" is well known from standard Hartree–Fock schemes and automatically accounts for all two-body forces in the ansatz. The density-dependent term ($\propto t_3$) additionally requires the so-called rearrangement energy, E_{rearr}. The density-dependent approximation to the Coulomb exchange also contributes to the rearrangement energy as discussed in the next subsection. The final energy is computed in the code according to (2.13–2.15).

2.2.2   The Coulomb Energy

The Coulomb interaction is a well-known piece of the nuclear interaction. However, its infinite range makes it very time consuming to evaluate the exchange part exactly, and it is unwise to spend most of the computing time on a small contribution. Therefore the Coulomb-exchange part is treated in the so-called Slater approximation, and we obtain for the Coulomb energy

$$E_{\text{Coul}} = \frac{1}{2}e^2 \int d^3r d^3r' \rho_C(\mathbf{r}) \frac{1}{|\mathbf{r}-\mathbf{r}'|}\rho_C(\mathbf{r}') + E_{\text{Coul,exch}} \tag{2.16}$$

$$E_{\text{Coul,exch}} = -\frac{3}{4}\left(\frac{3}{\pi}\right)^{1/3} 4\pi \int_0^\infty dr\, r^2 \rho_{\text{pr}}^{4/3} . \tag{2.17}$$

The contribution to the Hartree–Fock potential is easily computed by variation:

$$U_{\text{Coul}} = U_{\text{Coul,dir}} + U_{\text{Coul,exch}} , \tag{2.18}$$

$$-\Delta U_{\text{Coul,dir}} = 4\pi e^2 \rho_C , \tag{2.19}$$

$$U_{\text{Coul,exch}} = -\left(\frac{3}{\pi}\right)^{1/3} \rho_{\text{pr}}^{1/3} , \tag{2.20}$$

where the direct part is determined from solving the radial Poisson equation (2.19) rather than by integration. The density-dependent approach to the exchange term requires a contribution to the rearrangment energy

$$E_{\text{Coul,rearr}} = \frac{1}{4}\left(\frac{3}{\pi}\right)^{1/3} 4\pi \int_0^\infty dr\, r^2 \rho_{\text{pr}}^{4/3} .$$ (2.21)

In the direct term, the nuclear charge distribution ρ_C should enter. It should include folding with the finite size of the proton, but this folding again is a bit slow, and one would like to avoid it during the many repeated steps of an iteration. To a very good approximation one can replace $\rho_C \rightarrow \rho_{\text{pr}}$ in the Coulomb energy (2.16). This approximation is toggled by IFRHOC in the code (see Sect. 2.6.2).

There are many applications in the literature which do not include the Coulomb-exchange part at all. We allow for this in the code with the switch IFEX, as explained in Sect. 2.6.2.

2.2.3 Pairing

A schematic pairing force is introduced through the energy functional

$$E_{\text{pair}} = -\sum_q G_q \left[\sum_{\beta \in q} \sqrt{w_\beta(1 - w_\beta)}\right]^2 ,$$ (2.22)

where the pairing matrix elements G_q are constant within each species, $q \in \{\text{pr}, \text{ne}\}$. The BCS equations for the pairing weights w_β are obtained by varying the energy functional (2.1) with respect to w_β. This yields the standard BCS equations for the case of a constant pairing force. The occupation weights become

$$w_\beta = \frac{1}{2}\left(1 - \frac{\epsilon_\beta - \epsilon_{F,q}}{\sqrt{(\epsilon_\beta - \epsilon_{F,q})^2 + \Delta_q^2}}\right) .$$ (2.23)

The two remaining parameters, the pairing gap Δ_q and the Fermi energy $\epsilon_{F,q}$, are determined by simultaneous solution of the gap equation and the particle-number condition:

$$\frac{\Delta_q}{G_q} = \sum_{\beta \in q} \sqrt{w_\beta(1 - w_\beta)} ,$$ (2.24)

$$A_q = \sum_{\beta \in q} w_\beta ,$$ (2.25)

where A_q is the desired number of protons ($q = \text{pr}$) or neutrons ($q = \text{ne}$). In the following we call this treatment the *constant-force approach*.

Often one simplifies the pairing treatment further by parametrizing the pairing gap Δ_q directly [2.10]. In this case only the Fermi energy $\epsilon_{F,q}$ has to be adjusted such that the particle-number condition (2.25) is matched, and the gap equation (2.24) need not be solved. It does, however, serve to compute G_q, which then is needed to compute the pairing energy according to (2.22). In the following we call this procedure the *constant-gap approach*. A commonly accepted parametrization of the gap is

$$\Delta_q = (11.2 \text{ MeV})/\sqrt{A}\,, \tag{2.26}$$

where $A = A_{\text{pr}} + A_{\text{ne}}$ is the total nucleon number of a nucleons.

Both variants of the pairing treatment are implemented in the code. They are switched by the sign of the pairing strength in the input data; for details see Sect. 2.6.2.

2.2.4 The Center-of-Mass Correction

The exact nuclear ground state should be a state with total momentum zero. The mean field, however, localizes the nucleus and thus breaks translational invariance [2.11], so that the center-of-mass of the whole nucleus oscillates in the mean-field. In principle, one should project a state with good total momentum zero out of the given mean-field state. A simple and reliable substitute for the projection is to subtract the zero-point energy of the nearly harmonic oscillations of the center of mass [2.7]. The correction is simply

$$E_{\text{cm}} = \frac{\langle P_{\text{cm}}^2 \rangle}{2Am}\,, \tag{2.27}$$

$$\langle P_{\text{cm}}^2 \rangle = \sum_{\beta} w_{\beta} \langle \varphi_{\beta} \mid \hat{p}^2 \mid \varphi_{\beta} \rangle$$

$$- \sum_{\alpha,\beta} \left(w_{\alpha} w_{\beta} + \sqrt{w_{\alpha}(1 - w_{\alpha}) w_{\beta}(1 - w_{\beta})} \right) \mid \langle \varphi_{\alpha} \mid \hat{p} \mid \varphi_{\beta} \rangle \mid^2\,,$$

where $P_{\text{cm}} = \sum_i \hat{p}_i$ is the total momentum operator, A the nucleon number, and m the average nucleon mass. Note that this means evaluating a two-body operator, which leads to a double-sum over single-particle states. It would be too time consuming to evaluate such a small but involved correction at each iteration. Thus the center-of-mass correction (2.27) is computed *after variation*, and does not contribute to the Hartree–Fock equations (2.8–2.12). The correction is subtracted only a posteriori.

An even simpler approach to the center-of-mass correction was used in early applications. One can take the one-body part of (2.27), $P_{\text{cm}}^2 \approx \sum_i \hat{p}_i^2$, and thus express the center-of-mass correction as a modification of the nucleon mass

$$\frac{\hbar^2}{2m} \rightarrow \frac{\hbar^2}{2m} \left(1 - \frac{1}{A} \right)\,. \tag{2.28}$$

Both variants are implemented in the code. They are switched by IFZPE; see Sect. 2.6.2.

2.3   Representation of Wave Functions and Fields

2.3.1   The Radial Grid

After the spherical separation (2.4), the radial wave functions R_β and the densities and fields ρ_q, τ_q, etc. depend only on the radial distance r. We represent them on an equidistant radial grid

$$f(r) \equiv f_i = f(r_i), \quad r_i = (i-1)\Delta_r \quad , \quad i = 1, ..., N_r , \tag{2.29}$$

where f stands for any one of the wave functions or fields.

The radial integration is done by the simple trapezoidal rule:

$$\int_0^\infty dr\, F(r) \approx \Delta_r \left(\tfrac{1}{2}F_1 + \sum_{i=2}^{N_r-1} F_i + \tfrac{1}{2}F_{N_r} \right) = \Delta_r \sum_{i=2}^{N_r} F_i , \tag{2.30}$$

where we have used the fact that all integrands that occur in spherical calculations vanish at the origin $(i = 1)$ and are negligible at the boundary $(i = N_r)$. We have found out in tests that the simple trapezoidal rule is very precise for integrands that vanish at the boundaries, comparable to or better than a five-point integration formula.

For the derivatives, we use five point-precision throughout. This implies

$$(\partial_r f)_i \;=\; \frac{1}{\Delta_r}(-\tfrac{1}{12}f_{i+2} + \tfrac{2}{3}f_{i+1} - \tfrac{2}{3}f_{i-1} + \tfrac{1}{12}f_{i-2}) , \tag{2.31}$$

$$(\partial_r^2 f)_i \;=\; \frac{1}{\Delta_r^2}(-\tfrac{1}{12}f_{i+2} + \tfrac{4}{3}f_{i+1} - \tfrac{5}{2}f_i + \tfrac{4}{3}f_{i-1} - \tfrac{1}{12}f_{i-2}) . \tag{2.32}$$

These formulas cease to work at the boundaries of the grid because they call the function f at points outside the grid, e.g. f_0. Fortunately, all functions have well-defined reflection symmetry about the lower boundary $r = 0$. The radial wave functions are $R_\beta \propto r^{l_\beta+1} P(r^2)$, where P can be any polynomial, and the densities and potentials are even, i.e. $\propto P(r^2)$. Thus we can define lower grid values by $f_0 = \Pi f_2$ and $f_{-i} = \Pi f_{i+2}$ with Π the radial parity of the field f. The solution of the boundary problem is less clean at the upper bound, but all wave functions and fields should be very small there. Thus, one may very well switch to a less precise treatment. We use three-point precision for the derivatives at the penultimate point, $i = N_r - 1$, and extrapolate to the upper boundary by $\mathcal{F}_{N_r} = 2\mathcal{F}_{N_r-1} - \mathcal{F}_{N_r-2}$ with $\mathcal{F}$ standing for $\partial_r f$ or $\partial_r^2 f$. A special treatment is used for the second derivative in the inversion of the mean-field Hamiltonian, as will be discussed in connection with the inverse gradient step in Sect. 2.4.

2.3.2   Storage of Densities and Potentials

All densities (ρ_q, τ_q, J_q) and potentials ($\mathcal{B}_q, U_q, U_{ls,q}$) occur once for protons ($q = \mathrm{pr}$) and once for neutrons ($q = \mathrm{ne}$). We define one linear array for each field, e.g. RHO for the density ρ, and stack the neutron array above the proton array, so that

$$\begin{aligned}
\rho_{\text{pr}} &\leftrightarrow \text{RHO}(1, ..., N_r) \,, \\
\rho_{\text{ne}} &\leftrightarrow \text{RHO}(N_r + 1, ..., 2N_r) \,,
\end{aligned} \qquad (2.33)$$

and similarly for all other fields.

2.3.3   Shell Ordering

The quantum number β for the single-particle states is, in fact, a composite quantum number, $\beta = (q, n_\beta, j_\beta, l_\beta, m_\beta)$. While the m_β is irrelevant for radial properties, there still remains a multiplet of four quantum numbers, which run over very different values. The typical arrangement in nuclear-shell models is described very well by the ordering of shells as given in the harmonic-oscillator shell model [2.12], as shown in Table 2.1. It is advantageous to pack the total angular momentum j and the orbital angular momentum l into one combined quantum number

$$\text{JS} = j + l + \tfrac{1}{2} \,, \qquad (2.34)$$

which in turn allows retrieving the separate angular momenta as $l = \left[\frac{\text{JS}}{2}\right]$ and $j = \left[\frac{\text{JS}+1}{2}\right] - \frac{1}{2}$, where [...] means the "integer part".

We adopt the oscillator shell ordering as given in Table 2.1 for our book-keeping of states in the code. The single-particle states β are labeled by the shell number NSHL, and the single-particle energies ϵ_β and the occupation weights w_β are stored in a linear array according to this ordering. Again, we store proton and neutron information in *one* linear array, and the array of neutron information is stacked above the proton information.

There are two ways to loop through the single-particle states in the code. First, one can count along the shell number NSHL. The associated angular momentum JS is then obtained from the bookkeeping field JSP(NSHL). Alternatively, one can do a double loop over JS and the radial quantum number n. In this case one needs the reverse pointer that produces the NSHL for given JS and n. This is provided in the pointer field NPLA, which is, in fact, a pointer for the wave functions. We refer to the next subsection for more details.

2.3.4   Storage of Wave Functions

The radial wave functions R_β are stored in one linear array WF in portions of length N_r. They are stacked above one another according to the oscillator shell ordering of Table 2.1. The neutron wave functions follow on top of the proton wave functions, so that the arrangement becomes

Table 2.1. Typical shell ordering according to the nuclear oscillator shell model. This holds for protons as well as for neutrons. Shell closures are indicated by a horizontal line. The ordering of states within two shell closures may differ from model to model. For simple bookkeeping we have chosen a subordering by decreasing the total angular momentum j. The "degeneracy" is the number of nucleons fitting in the shell, and "accumulated A" denotes the number of nucleons up to and including the listed shell, thus immediately displaying the magic nucleon numbers at the shell closures.

NSHL	n	j	l	JS	name	degeneracy	accumulated A
1	1	1/2	0	1	$1s\frac{1}{2}$	2	2
2	1	3/2	1	3	$1p\frac{3}{2}$	4	6
3	1	1/2	1	2	$1p\frac{1}{2}$	2	8
4	1	5/2	2	5	$1d\frac{5}{2}$	6	14
5	1	3/2	2	4	$1d\frac{3}{2}$	4	18
6	2	1/2	0	1	$2s\frac{1}{2}$	2	20
7	1	7/2	3	7	$1f\frac{7}{2}$	8	28
8	1	5/2	3	6	$1f\frac{5}{2}$	6	34
9	2	3/2	1	3	$2p\frac{3}{2}$	4	38
10	2	1/2	1	2	$2p\frac{1}{2}$	2	40
11	1	9/2	4	9	$1g\frac{9}{2}$	10	50
12	1	7/2	4	8	$1g\frac{7}{2}$	8	58
13	2	5/2	2	5	$2d\frac{5}{2}$	6	64
14	2	3/2	2	4	$2d\frac{3}{2}$	4	68
15	3	1/2	0	1	$3s\frac{1}{2}$	2	70
16	1	11/2	5	11	$1h\frac{11}{2}$	12	82
17	1	9/2	5	10	$1h\frac{9}{2}$	10	92
18	2	7/2	3	7	$2f\frac{7}{2}$	8	100
19	2	5/2	3	6	$2f\frac{5}{2}$	6	106
20	3	3/2	1	3	$3p\frac{3}{2}$	4	110
21	3	1/2	1	2	$3p\frac{1}{2}$	2	112
22	1	13/2	6	13	$1i\frac{13}{2}$	14	126
23	1	11/2	6	12	$1i\frac{11}{2}$	12	138
24	2	9/2	4	9	$2g\frac{9}{2}$	10	148
25	2	7/2	4	8	$2g\frac{7}{2}$	8	156
26	3	5/2	2	5	$4d\frac{5}{2}$	6	162
27	3	3/2	2	4	$4d\frac{3}{2}$	4	166
28	4	1/2	0	1	$5s\frac{1}{2}$	2	168
29	1	15/2	7	15	$1j\frac{15}{2}$	16	184

$$
\begin{aligned}
R_{\mathrm{pr},1\mathrm{s}\frac{1}{2}} &\equiv \mathrm{WF}(1,...,N_r) \\
R_{\mathrm{pr},1\mathrm{p}\frac{3}{2}} &\equiv \mathrm{WF}(N_r+1,...,2N_r) \\
R_{\mathrm{pr},1\mathrm{p}\frac{1}{2}} &\equiv \mathrm{WF}(2N_r+1,...,3N_r) \\
... \qquad &\quad ... \\
R_{\mathrm{ne},1\mathrm{s}\frac{1}{2}} &\equiv \mathrm{WF}(\mathcal{A}+1,...,\mathcal{A}+N_r) \\
R_{\mathrm{ne},1\mathrm{p}\frac{3}{2}} &\equiv \mathrm{WF}(\mathcal{A}+N_r+1,...,\mathcal{A}+2N_r) \\
... \qquad &\quad ...
\end{aligned}
\qquad\qquad (2.35)
$$

where $\mathcal{A} = N_r M_{\mathrm{pr}}$ is the offset for neutron storage and $M_{\mathrm{pr}} =$ "number of proton shells". The first position in WF for a wave function for a given state (q,n,j,l) is stored in the pointer field NPLA. It can easily be constructed from the scheme (2.35) with the shell ordering of Table 2.1 as $\mathrm{NPLA} = N_r * \mathrm{NSHL} + 1$ for protons and $\mathrm{NPLA} = \mathcal{A} + N_r * \mathrm{NSHL} + 1$ for neutrons. Note that the pointer field can also be used to access the number of a state in the fields for the ϵ_β and for the w_β: the integer part of NPLA/N_r returns just NSHL with the appropriate offset.

2.4 Iterative Solution of the Hartree–Fock Equations

2.4.1 The Gradient Iteration

The Hartree–Fock equations (2.8–12) are nonlinear in the wave functions, so that an iterative solution is needed. It is advisable then also to use an iterative scheme for solving the differential equation (2.8) because the intermediate solutions for the wave functions are superseded anyway. The simplest and most robust iteration scheme to find an eigenvalue of h is $\varphi_\beta^{(n+1)} = h\varphi_\beta^{(n)}$. This scheme will converge towards $h_{\max}$, the maximum eigenvalue of h. However, we are interested rather in the opposite, the minimum eigenvalue $h_{\min}$. To this end, we construct from h a step-operator $\mathcal{S}$ such that $h_{\min}$ corresponds to $\mathcal{S}_{\max}$. This yields the gradient iteration

$$
\varphi_\beta^{(n+1)} = \mathcal{O}\{\mathcal{S}^{(n)}\varphi_\beta^{(n)}\} , \qquad\qquad (2.36)
$$

where $\mathcal{O}$ denotes orthonormalization of all φ_β in ascending order of the single-particle states β and where the (simple) gradient step is

$$
\mathcal{S}_{\mathrm{grad}}^{(n)} = 1 - \delta h^{(n)} , \quad \text{where} \quad \delta < \frac{2}{h_{\max}} . \qquad\qquad (2.37)
$$

The restriction on δ guarantees that $h_{\min}$ produces $\mathcal{S}_{\max}$. Note that $h^{(n)}$ and subsequently $\mathcal{S}_{\mathrm{grad}}^{(n)}$ depend on the iteration because the Hartree–Fock equations are nonlinear equations in $\varphi_\beta^{(n)}$. The step (2.37) is called the (simple) gradient step because $h\varphi_\beta$ is used to approach the goal and this is just the gradient of the potential-energy landscape, $h\varphi_\beta = \partial E/\partial\varphi_\beta^+$.

The gradient step is simple and robust, but can become extremely slow owing to large h_{max} in a good numerical representation of the φ_β. For example with the kinetic energy in a finite-difference scheme with five-point precision we obtain $h_{max} \approx \frac{\hbar^2}{2m\Delta r^2}(\frac{16}{3} + l_{max}(l_{max} + 1))$, which ranges from 2000 MeV to 5000 MeV in typical nuclear applications, where we are looking for h_{min} in the range $10 - 50$ MeV. Thus, one typically needs a few thousand iterations to obtain satisfactory precision from the simple gradient iteration.

2.4.2 The Damped Gradient Iteration

The problem with the large h_{max} and the resulting slow convergence of the gradient step can easily be localized for the case of the nuclear shell model: it is the kinetic energy which rises to annoyingly large values of thousands of MeV, whereas the potential energy typically stays within 100 MeV. This suggests improving the step by damping the components with high kinetic energy and leads to the damped gradient step

$$ \mathcal{S}^{(n)\text{damp}} = 1 - \delta \left(\frac{p^2}{2m} + E_0 \right)^{-1} \left(h^{(n)} - \langle \varphi_\beta^{(n)} \mid h^{(n)} \mid \varphi_\beta^{(n)} \rangle \right) , \qquad (2.38) $$

where δ and E_0 are numerical parameters to be adjusted for optimal speed and stability. Typical values are $\delta \approx 1$ and $E_0 \approx 40$ MeV, which is about the variation of the potential. The subtraction of the expectation value $\langle h \rangle$ is a shortcut to project the step orthogonal to the solution already reached; it guarantees that the step decays properly as one approaches the solution.

The damped gradient step (2.38) is a substantial improvement over the simple gradient step. One usually obtains five-digit precision within about 30–50 iterations. The method has been applied successfully in many different problems with different representations of the fields (finite differences, fast Fourier, B-splines) and different dimensions (spherical 1D, axially symmetric 2D, full Cartesian 3D) [2.6, 7, 13]. For applications in more than one dimension, one can reduce the potentially time-consuming inversion of the kinetic energy by a separable approach, $(1 + \frac{p^2}{2mE_0}) \approx \prod_{i=x,y,z}(1 + \frac{p_i^2}{2mE_0})$. Thus one only needs successive inversions in one dimension, which makes the damped gradient an extremely fast scheme in higher dimensions.

2.4.3 The Inverse Gradient Step

In one-dimensional calculations, it is even conceivable to invert the full mean-field Hamiltonian h, that is, replace $\frac{p^2}{2m}$ in the damped gradient step (2.38) by h. One finds that optimum convergence is achieved for parameter combinations δ and E_0 such that the step becomes a mere inversion of h. This leads to the inverse gradient step

$$ \mathcal{S}_{\text{inv}}^{(n)} = \left(h^{(n)} - U_{\text{min}}^{(n)} \right)^{-1} , \qquad (2.39) $$

where $U_{\min}^{(n)}$ is the minimum value of the mean-field potential. $U_{\min}$ is needed to make the inverted operator positive definite. This step is even a bit faster than the damped gradient step. An additional advantage is that it requires fewer operations per iteration and that it is free of numerical parameters. It should be noted that the step may be made even faster by using $\epsilon_\beta^{(n)} + \delta$ rather than $U_{\min}^{(n)}$. But this is potentially dangerous and can lead to instabilities.

The inverse gradient step (2.39) is our preferred step for spherical calculations of the nuclear ground state. We nevertheless provide both steps, the inverse gradient step and the damped gradient step, in the code. This gives the user the opportunity to get experience with both schemes. The switch for the steps in the code is the stepsize $\delta = \text{X0DAMP}$: the inverse gradient step (2.39) is used for $\text{X0DAMP} = 0$ whereas a finite value switches to the damped gradient step (2.38).

The wave-function iteration with the inverse gradient step is so fast that the change in the potential fields gives rise to unwanted oscillations. We damp them by mixing only a fraction ADDNEW of the new potential $U^{(n)}$ with the previous potential $U^{(n-1)}$.

2.4.4 Efficient Inversion with Five-Point Precision

The five-point formula (2.32) for the second derivative makes the kinetic energy a sparse matrix with two off-diagonal lines. This could easily and quickly be inverted by a Gaussian elimination scheme. However, there is an even faster scheme [2.14]. We write the five-point formula for the second derivative as

$$
\begin{aligned}
\Delta_5 &= D_3^{-1}\Delta_3 \,, \\
(D_3 f)_i &= \tfrac{1}{12}f_{i+1} + \tfrac{5}{6}f_i + \tfrac{1}{12}f_{i-1} \\
(\Delta_3 f)_i &= f_{i+1} - 2f_i + f_{i-1} \,.
\end{aligned}
\tag{2.40}
$$

The inversion is written as an inhomogenous linear equation. For the kinetic energy inversion (e.g., $\psi = (\frac{p^2}{2m} + E_0)^{-1}\varphi$), we write

$$
\left(\frac{p^2}{2m} + E_0\right)\psi = \varphi \,,
$$

to be solved for ψ. We insert the second derivative in the approach (2.40) and operate with D_3 from left. Thus we have to solve

$$
\left(-\frac{\hbar^2}{2m}\Delta_3 + E_0 D_3\right)\varphi = D_3\psi \,.
\tag{2.41}
$$

The matrix to be inverted is now sparse with only one off-diagonal line. This, of course, can be inverted even faster by a Gaussian elimination scheme. Thus we obtain five-point precision in nearly the same time as three-point precision.

2.4.5 Pairing Iteration

The pairing equations are solved by Newton's tangential iteration. We illustrate it for the "constant-gap approach", where we have to determine the Fermi energy ϵ_F such that the particle-number condition (2.25) is met for a given set of single-particle energies ϵ_β. There may be a deviation from the goal for some given Fermi energy $\epsilon_{F,q}^{(n)}$. We linearize the error and determine a correction $\delta\epsilon_{F,q}$ by

$$A_q = \sum_{\beta \in q} w_\beta + \sum_{\beta \in q} \frac{\partial w_\beta}{\partial \epsilon_{,q}} \delta\epsilon_{F,q} , \qquad (2.42)$$

such that the new $\epsilon_{F,q}^{(n+1)} = \epsilon_{F,q}^{(n)} + \delta\epsilon_{F,q}$ comes much closer to the proper Fermi energy. The derivative in (2.42) is determined numerically. As the tangential step has a tendency to overshoot, we stabilize it by reducing it to 90%. Thus, we iterate

$$\epsilon_{F,q}^{(n+1)} = \epsilon_{F,q}^{(n)} + 0.9 \frac{A_q - \sum_{\beta \in q} w_\beta}{\sum_{\beta \in q} \frac{\partial w_\beta}{\partial \epsilon_{F,q}}} . \qquad (2.43)$$

This step still suffers in that it sometimes shoots completely out of a physically reasonable region. As a further safety measure, we limit the step in $\epsilon_{F,q}$ to a maximum change of $3\,\mathrm{MeV}$.

The iteration for the "constant-force approach" proceeds very similarly. We then have to compute two steps, $\delta\epsilon_{F,q}$ and $\delta\Delta_q$, from a twofold variation of the two equations (2.24) and (2.25). The extension is obvious and will not be outlined in detail here.

The Hartree–Fock and pairing equations are coupled, and they are solved by simultanous iteration of the wave functions and the occupation weights. The procedure is outlined in Sect. 2.6.

2.5 The Form Factor and Related Observables

2.5.1 Composing the Nuclear Charge Density

The nuclear charge density is a most useful observable for analyzing nuclear structure: it provides information about the nuclear shape and can be determined by clear-cut procedures from the cross section for elastic electron scattering [2.15]. To compute the observable charge density from the Hartree–Fock results, one has to take into account that the nucleons themselves have an intrinsic electromagnetic structure [2.16]. Thus one needs to fold the proton and neutron densities from the Hartree–Fock method with the intrinsic charge density of the nucleons. Folding becomes a simple product in Fourier space, so that we transform the densities to the so-called form factors

$$F_q(k) = 4\pi \int_0^\infty dr\, r^2 j_0(kr)\rho_q(r) , \qquad (2.44)$$

Table 2.2. The coefficients of the isospin-coupled Sachs form factors of the nucleons; the b_i are given in fm^{-1}.

	a_1	a_2	a_3	a_4	b_1	b_2	b_3	b_4
$E, I = 0$	2.2907	-0.6777	-0.7923	0.1793	15.75	26.68	41.04	134.2
$E, I = 1$	0.3681	1.2263	-0.6316	0.0372	5.00	15.02	44.08	154.2
M	0.6940	0.7190	-0.4180	0.0050	8.50	15.02	44.08	355.4

where j_0 is the spherical Bessel function of zeroth order. In fact, the form factor is closer to experiment because in the Born approximation it directly represents the amplitude for scattering at momentum transfer $\hbar k$. The charge form factor is then given by

$$F_{\mathrm{C}}(k) = \sum_q [F_q(k) G_{\mathrm{E},q}(k) + F_{ls,q}(k) G_{\mathrm{M}}(k)] \exp\left(\tfrac{1}{8}(\hbar k)^2 / \langle P_{\mathrm{cm}}^2 \rangle\right), \qquad (2.45)$$

where $F_{ls,q}$ is the form factor of the spin–orbit current $\nabla \boldsymbol{J}$ (accounting for magnetic contributions to the charge density), $G_{\mathrm{E},q}$ are the electric form factors of the nucleons, G_{M} is the magnetic form factor of the nucleons (assumed to be equal for both species), and the exponential factor takes into account an unfolding of the spurious vibrations of the nuclear center of mass in the harmonic approximation (the $\langle P_{\mathrm{cm}}^2 \rangle$ therein is the same as in the zero-point energy (2.27)).

The intrinsic nucleon form factors are taken from electron scattering on protons and deuterons. They are given as the so-called Sachs form factors for isospin 0 and 1 in the form of a dipole fit, [2.17]

$$G_s = \sum_{\nu=1}^{4} \frac{a_{s,\nu}}{1 + k^2/b_{s,\nu}} \quad \text{for } s \in \{(\mathrm{E}, I = 0), (\mathrm{E}, I = 1), \mathrm{M}\} . \qquad (2.46)$$

The actual coefficients are displayed in Table 2.2. The isospin-coupled form factors are recoupled for our purposes to proton and neutron form factors, and the relativistic Darwin correction [2.16] has to be added, yielding

$$G_{\mathrm{E},pr/ne} = \tfrac{1}{2}(G_{\mathrm{E},I=0} \pm G_{\mathrm{E},I=1}) / \sqrt{1 + \frac{\hbar^2 k^2}{2m_{pr/ne}}} . \qquad (2.47)$$

Note that the nucleon structure is taken into account only approximately because we fold with the free form factors of the nucleons thus neglecting medium distortion and off-shell effects of the nucleon.

The charge density is obtained from the charge form factor by the inverse Fourier–Bessel transform

$$\rho_{\mathrm{C}}(r) = \frac{1}{2\pi^2} \int \mathrm{d}k\, k^2 j_0(kr) F_{\mathrm{C}}(k) . \qquad (2.48)$$

Other information can be drawn directly from the form factor, as we will see in the next subsection.

2.5.2 Radii and Surface Thickness

Most of the information that is contained in the form factor at low q can be described by two parameters: the diffraction radius R and the surface thickness σ [2.18]. The diffraction radius is determined from the first zero of the form factor

$$R = 4.493/k_0^{(1)} \quad \text{where} \quad F_C(k_0^{(1)}) = 0 . \tag{2.49}$$

It parametrizes the overall diffraction pattern, which resembles that of a box with radius R. The diffraction radius is the box equivalent radius.

Upon closer inspection one realises that the nuclear form factor decreases more rapidly than the box form factor. This is due to the finite surface thickness of nuclei. If one models the nuclear surface by folding the box distribution with a Gaussian, $\exp\left(-\frac{1}{2}q^2\sigma^2\right)$, one can determine the surface parameter σ by comparing the heights of the first maxima of the box-equivalent and the Hartree–Fock results:

$$\sigma = \frac{2}{k_m} \log\left(F_{box}(k_m)/F_C(k_m)\right) \quad \text{with} \quad F_{box}(k) = 3\frac{j_1(k_m R)}{k_m R} , \tag{2.50}$$

where $k_m = 5.6/R$ is the momentum where F_{box} has its first maximum.

Alternative information on the overall nuclear size is given by the root-mean-square (r.m.s.) radius r, which is determined from the curvature of the form factor at $k \to 0$ as

$$r = 3\left(\frac{d^2}{dk^2}F_C(k)|_{k=0}\right)/F_C(0) . \tag{2.51}$$

It does not carry much new information because it can be related to R and σ approximately as $r = \sqrt{R^2 + 3\sigma^2}$ [2.18]. The parametrization in terms of R and σ is preferable because the diffraction radius R is very stable against correlations and depends smoothly on the particle number A, whereas the surface thickness σ is sensitive to shell effects and correlations from low energy modes.

2.5.3 Handling Form Factors on a Grid

In practice, the nucleon densities are given on a finite radial grid (see Sect. 2.3). All information on densities is retained if one computes and stores the form factor on the reciprocal lattice,

$$F(k) \equiv F(k_j) \quad , \quad k_j = (j-1)\frac{\pi}{\Delta_r N_r} \quad , \quad j = 1, ..., N_k , \tag{2.52}$$

where usually $N_k = N_r$. In practice, one chooses the grid in coordinate space finer than necessary for a grid in momentum space since the finite difference formulas for the kinetic energy are inferior to a Fourier representation. Thus, one can easily drop the high Fourier components and choose $N_k \approx \frac{1}{2}N_r$.

The Fourier-Bessel forward and inverse transformations (2.44) and (2.48) are evaluated by the simple trapezoidal rule, which corresponds to the usual method of Fourier transformations between reciprocal lattices.

The form factor is thus given on a grid $\{k_j, j = 1, ..., N_k\}$. But all three form parameters defined in the previous subsection require access to the form factor F_C at any value of k. We compute these intermediate values by Fourier–Bessel interpolation

$$F(k) = \frac{\sin(kR_{max})}{kR_{max}} \sum_{j=1}^{N_k} (-)^{j+1} \left[1 - \frac{kR_{max}}{j\pi}\right]^{-1} F(k_j) , \qquad (2.53)$$

where $R_{max} = N_r \Delta_r$ is the length of the coordinate grid. The interpolation runs into problems with round-off errors near $k \approx j\pi/R_{max}$; we switch to a Taylor expansion in powers of $k - k_j$ near those critical points.

2.6   Instructions and Suggestions

At this point, papers usually have conclusions. For a presentation of a code, the appropriate end is an invitation to use the code and to experiment with it. Therefore, in this final section we give a short overview of the code structure and of the necessary input.

2.6.1   Structure of the Code

The main program, HAFOMN, serves only as a short master routine that sets default values for some of the parameters, contains all READ statements, and then quickly calls the principal subroutine HAFOSU which performs all other tasks, such as initialization, iteration, final analysis, and printing. This strategy to delegate everything except reading to a "master subroutine" was chosen to make the whole Hartree–Fock package easily callable from other routines. We used it, for example, extensively as the core of a least-squares fitting routine. All transfer to and from HAFOSU is done via common blocks. For detailed information on the I/O see the comments at the beginning of HAFOSU. A full explanation of each variable in each of the COMMON blocks is given at the top of the master routine HAFOMN.

The coupled Hartree–Fock and pairing equations are solved by an interleaved simultaneous iteration, preceded by an initialization phase and followed by an analysis (and output) phase. The structure of the code within HAFOSU is shown diagrammatically in Fig. 2.1. The decision to iterate a few pairing steps per wave-function step seems to introduce some asymmetry. The code under most circumstances also works with one pairing step per iteration. But the pairing step is orders of magnitude faster than the wave-function step. Thus, it is not worthwhile to be sparing with pairing steps at the (even small) risk of slowing down the whole iteration.

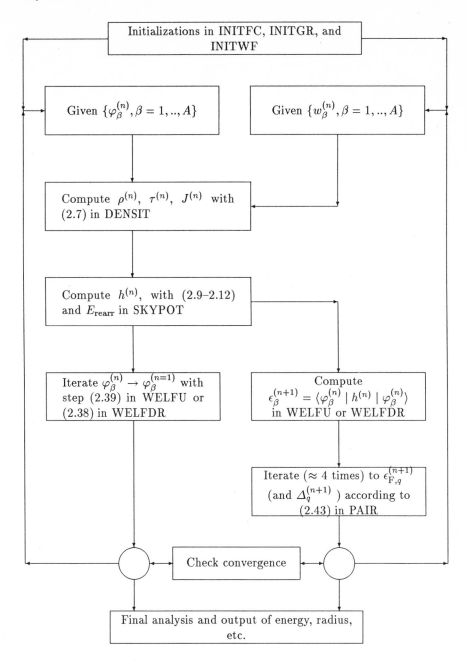

Fig. 2.1. Schematic flow chart of the code.

2.6.2 Examples for Input Data

In Table 2.3, we show a typical set of input lines to be placed into the file
FOR005. Although all input variables are explained in the code, here we briefly
discuss the input lines. We give input for three successive calculations with
varying options and varying forces. The first set uses the force "Fit Z_σ", the
second set "Skyrme M*", and the third set the old fashioned "Skyrme 3"; for
definitions, references, and a variety of other parametrizations see Ref. [2.3].
In the first set, we use default input parameters wherever possible by entering
no value at those places; the double magic nucleus ^{208}Pb is calculated. In the
second set, all parameters are given explicitly but still with their default
values; ^{124}Sn is computed. It has a magic proton number but needs pairing
for the neutrons. Note that the low ITPR will produce error messages from
the pairing routine, PAIR, during the first few iterations. One should not be
worried by this because pairing converges well in the later iterations. In the
third set, we have varied some options. Note that the pairing forces PAIRFP
and PAIRFN are set to 0, which switches off the pairing iteration and calls
for further input, namely the lines which explicitly specify the occupation
weights w_β. The nucleus ^{17}O is computed here in the filling approximation.
All three sets start with IWUNIT = 0, which will produce output in file
FOR006. In the following, we briefly explain each input variable:

IWUNIT Determines the output unit ($0 \rightarrow$ file FOR006, $1 \rightarrow$ terminal)
or terminates the program (-1).

NGRID = N_r = number of radial grid points. If NGRID = 0 is entered,
the program automatically computes a reasonable grid size according
to the number of nucleons (assuming normal nuclear sizes). *Default =
0.*

RSTEP = Δ_r = grid spacing in units of fm. *Default = 0.3*, which suffices
for 0.1% precision in energy and radius.

IWR = print level. Possible values are $-2, -1, 0, 1$. The higher the value
the more detailed the printed output. *Default = -1.*

IFPLOT = switch for output on a plot file FOR011 (*Default = 0*):
0 $\rightarrow$ no plot file is written,
1 $\rightarrow$ a short plot file is written up to the single-particle energies,
2 $\rightarrow$ the densities are given in addition,
3 $\rightarrow$ the charge form factor is also given.

DABR = relative error in the radius which must be achieved before the
iteration terminates. *Default = 0.0001.*

DABE = relative error in the energy which must be achieved before the
iteration terminates. *Default = 0.0001.*

ITMAX = maximum number of iterations. *Default = 80*

EPSPR = precision of the pairing iteration.

ITPR = number of pairing steps per wave-function step. *Default = 4.*

Table 2.3. Example of input lines in file `FOR005`. The code reads only the numbers on the left half of the input lines. The right half is added as a comment to simplify changes in the input data.

```
0,                                          IWUNIT ''FIT Z-sigma ''
,,                                          NGRID,RSTEP
,,                                          IWR,IFPLOT
,,,,,                                       ITMAX,DABR,DABE,ITPR,EPSPR
,,,                                         XODAMP,EODAMP,ADDNEW
1,1,1,0,0,                                   IF:EX,TM,ZPE,DFMS,RHOC
-1983.76,362.252,-104.27,11861.4,123.69,    T0,T1,T2,T3,T4
1.1717,0.,0.,1.7620,                        X0,X1,X2,X3
0.25,1.0,,,                                 POWER,ACOUL,PAIRFP,PAIRFN
82,126,,,                                    NPROT,NNEUT,NMAXP,NMAXN
0,                                          IWUNIT ''Skyrme M*''
00,0.3,                                     NGRID,RSTEP
-1,0,                                       IWR,IFPLOT
80,0.0001,0.0001,4,0.0001,                   ITMAX,DABR,DABE,ITPR,EPSPR
0.0,40.0,0.5,                               XODAMP,EODAMP,ADDNEW
1,1,1,0,0,,                                  IF:EX,TM,ZPE,DFMS,RHOC
-2645.0,410.0,-135.0,15595.0,130.0,         T0,T1,T2,T3,T4
0.09,0.,0.,0.0,                             X0,X1,X2,X3
0.16666667,1.0,-1.1E11,-1.1E11,             POWER,ACOUL,GAPP,GAPN
50,74,00,00,                                 NPROT,NNEUT,NMAXP,NMAXN
0,                                          IWUNIT ''Skyrme 3 ''
00,0.3,                                     NGRID,RSTEP
-1,1,                                       IWR,IFPLOT
80,0.00001,0.00001,4,,                       ITMAX,DABR,DABE,ITPR,EPSPR
1.00,40.0,0.5,                              XODAMP,EODAMP,ADDNEW
1,0,0,0,0,                                   IF:EX,TM,ZPE,DFMS,RHOC
-1128.75,395.0,-95.0,14000.0,120.0,         T0,T1,T2,T3,T4
0.45,0.,0.,1.0,                             X0,X1,X2,X3
1.0,1.0,0.0,0.0,                            POWER,ACOUL,GAPP,GAPN
08,09,03,04,                                 NPROT,NNEUT,NMAXP,NMAXN
1,1,1,                                      WEIGHT(PROTONS)
1,1,1,.16666667,                            WEIGHT(NEUTRONS)
-1,                                         TO STOP
```

X0DAMP $= \delta =$ stepsize of damped gradient step (2.38). For $X0DAMP = 0$ the inverse gradient step (2.39) is run. *Default = 0.*

E0DAMP $= E_0 =$ energy parameter in kinetic-energy damping (2.38). *Default = 50.* Hint: try to explore optimum combinations of X0DAMP and E0DAMP; be aware that these may depend a bit on the force used.

ADDNEW $=$ weight for admixing the new potentials $U^{(n)}$ to the previous potentials $U^{(n-1)}$ to stabilize the iteration.

IFEX $=$ switch for Coulomb exchange according to (2.20) and (2.21):

  $1 \rightarrow$ exchange is taken into account,

  $0 \rightarrow$ exchange is omitted.

IFTM $=$ switch for t_1 and t_2 contribution in the spin-orbit potential $U_{ls,q}$:

  $0 \rightarrow$ the term is dropped,

  $1 \rightarrow$ the term is taken into account.

IFZPE $=$ switch for the handling of the center-of-mass correction:

  $1 \rightarrow$ full correction according to (2.27),

  $0 \rightarrow$ approximate correction according to (2.28).

IFDFMS $=$ switch for equal or different nucleon masses:

  $0 \rightarrow \frac{\hbar^2}{2m_{pr}} = \frac{\hbar^2}{2m_{ne}} = 20.7525\,\mathrm{MeV\,fm}^2$,

  $1 \rightarrow \frac{\hbar^2}{2m_{pr}} = 20.735\,\mathrm{MeV\,fm}^2$ and $\frac{\hbar^2}{2m_{ne}} = 20.721\,\mathrm{MeV\,fm}^2$.

IFRHOC $=$ switch for proton folding in Coulomb potential:

  $1 \rightarrow$ charge density is used in Coulomb potential (2.18),

  $0 \rightarrow$ proton density is used in Coulomb potential.

T0...X3 $=$ parameters of the Skyrme force, $t_0...x_3$.

POWER $= \alpha =$ parameter of the Skyrme force.

ACOUL $=$ fractional charge of the protons. *Default = 1.*

PAIRFP $=$ Pairing force for protons:

  $> 0 \rightarrow$ constant-force approach with $G_{pr} =$ PAIRFP,

  $> 10^{10} \rightarrow$ constant-force approach with $G_{pr} = 22\,\mathrm{MeV}/A$,

  $< 0 \rightarrow$ constant-gap approach with $\Delta_{pr} =$ ABS(PAIRFP),

  $< -10^{10} \rightarrow$ constant-gap approach with $\Delta_{pr} = 11.2\,\mathrm{MeV}/\sqrt{A}$,

  $= 0 \rightarrow$ override pairing and use WEIGHT from input line.

PAIRFN $=$ pairing force for neutrons. Switches as for PAIRFP, but uses $G_{ne} = 29\,\mathrm{MeV}/A$ in case of PAIRFN$> 10^{10}$.

NPROT $=$ number of protons.

NNEUT $=$ number of neutrons.

NMAXP $= M_{pr} =$ number of proton shells. The code automatically computes the appropriate NMAXP for given NPROT if NMAXP $= 0$ is entered. *Default = 0*

NMAXN $= M_{ne} =$ number of neutron shells. The code automatically computes the appropriate NMAXN for given NNEUT if NMAXN $= 0$ is entered. *Default = 0*

2.6.3 Technical Notes

The program has been tested on a PC with Prospero FORTRAN and RM-FORTRAN, on a CDC under NOS with the FTN5 compiler, and on an IBM 3090 with the VFORT compiler. One can just compile it and run. The program should also run immediately on a VAX. Slight differences between the machines occur only for the I/O. The assignment to the unit "CON" for direct output on the terminal (see the OPEN statement for UNIT=6 in HAFOMN) is specific for the PC and needs to be changed for each machine.

Each routine contains the same PARAMETER statement in the header. All PARAMETERs in the PARAMETER statement are explained in the header of the main program HAFOMN. All critical dimensioning is done via parameters. Thus the program can be easily readjusted for different circumstances by a global change of the appropriate parameter. We recommend collecting the PARAMETER statement and each COMMON block separately into an INCLUDE file and inserting the corresponding INCLUDE statement instead of the PARAMETER or COMMON sequence.

The code only uses generic names in function calls and never uses real constants in a calling list or an IF statement. Thus one can easily switch to double precision (although we rarely find it necessary). To that end each routine is headed by an IMPLICIT DOUBLE PRECISION which is switched off by a CDOUB in the first columns. One merely has to change all CDOUB into five blanks to convert the code to double precision.

References

[2.1] T.H.R. Skyrme, Nucl. Phys. **9** (1959) 615
       D. Vautherin and D. M. Brink, Phys. Rev. **C5** (1972) 626
[2.2] P. Quentin and H. Flocard, Ann. Rev. Nucl. Part. Sci. **28** (1978) 523
       K. Goeke and P.-G. Reinhard, Lecture Notes in Physics, vol **171** (1982)
[2.3] J. Friedrich and P.-G. Reinhard, Phys. Rev. **C33** (1986) 335
[2.4] H. Pfeiffer, P.-G. Reinhard, and D. Drechsel, Z. Phys. **A292** (1979) 375
[2.5] P.-G. Reinhard and R. Y. Cusson, Nucl. Phys. **A378** (1982) 418
[2.6] M. R. Strayer, R. Y.Cusson, A. S. Umar, P.-G. Reinhard, D. A. Bromley, and W. Greiner, Phys. Lett. **B135** (1984) 261
       R. Y. Cusson, P.-G. Reinhard, M. R. Strayer, J. A. Maruhn, and W. Greiner, Z. Phys. **A320** (1985) 475
[2.7] P.-G. Reinhard, F. Grümmer, and K. Goeke, Z. Phys. **A317** (1984) 339
[2.8] P.-G. Reinhard and J. Friedrich, Z. Phys. **321** (1985) 619
[2.9] A. R. Edmonds, *Angular Momentum in Quantum Mechanics* (Princeton University Press, Princeton 1957)
[2.10] J. Błocki and M. Flocard, Nucl. Phys. **A273** (1976) 45
[2.11] P. Ring and P. Schuck, *The Nuclear Many Body Problem* (Springer, Berlin, Heidelberg 1980

[2.12] J. M. Eisenberg and W. Greiner, *Nuclear Models* (North-Holland, Amsterdam 1971)

[2.13] S.-J. Lee et al., Phys. Rev. Lett. **57** (1986) 2916

[2.14] R. Zurmühl, *Praktische Mathematik* (Springer, Berlin, Heidelberg 1963) Chap. VII, Sect. 3

[2.15] B. Dreher, J. Friedrich, K. Merle, G. Lührs and H. Rothaas, Nucl. Phys. **A235** (1974) 219

[2.16] J. L. Friar, J. W. Negele, Adv. Nucl. Phys. **8** (1975) 219

[2.17] V. H. Walther, private communication; G. G. Simon, C. Schmitt, F. Borkowski, and V. H. Walther, Nucl. Phys. **A333** (1980) 318

[2.18] J. Friedrich and N. Voegeler, Nucl. Phys. **A373** (1982) 191

3. The Cranked Nilsson Model

T. Bengtsson, I. Ragnarsson and S. Åberg

3.1   Introduction

The Nilsson–Strutinsky method is presently the most feasible way to do systematic calculations of the nuclear energy as a function of deformation and/or angular momentum. These calculations are based on the developments of Bohr and Mottelson in the early 1950s [3.1,2] and the subsequent calculation of the nuclear single-particle orbitals as functions of deformation [3.3]. However, it was the methods of calculating the shell correction energy developed by Strutinsky [3.4,5] that made it possible to do large-scale calculations with some realism that describe, for example, the fission process. This led to a large number of calculations (e.g. [3.6–11]) where the nuclear potential-energy surface was explored in great detail as a function of different deformation degrees of freedom. Subsequently, the methods were developed to allow the angular-momentum degree of freedom to be explored within the same framework [3.12–15]. Important achievements in this field include the prediction of superdeformed high-spin states (e.g. [3.12–16]) and terminating bands; these phenomena were discussed extensively [3.17] long before they were observed experimentally.

We use here the modified single-particle oscillator potential. Frequently used alternatives are the Woods-Saxon [3.8,18] or Folded-Yukawa [3.7,19] potentials; those should be essentially equivalent from the physical point of view. The advantage of the modified oscillator is that all matrix elements can be given in analytic form. Therefore, compared with other potentials, it is often easier to extract the physical origin of new phenomena. Furthermore, numerical calculations of, for example, the shell correction energy become more stable and numerically reliable. The disadvantage lies in the potential's infinite walls, which imply that the wave functions are not very suitable in applications where the tails at large r-values are important. In addition, the ℓ^2 potential of the modified oscillator leads to an undesired velocity dependence, which shows up, for example, in the average moment of inertia [3.15]. In general, however, it is our experience that, with carefully chosen single-particle parameters, results using different potentials are very equivalent (see e.g. [3.20,21] or [3.22–3.25] for other references of a more general character.)

An alternative to average single-particle potentials are Hartree–Fock calculations [3.26,27]. These are, of course, very important even though phenomenological interactions are used in most cases. In general, however, such calculations are too lengthy to be used in large-scale systematic investigations. On the other hand they are very important for testing different approximations and assumptions and have thus helped to justify [3.28] the methods used in the present computer program.

This program has been developed to deal with quantities such as:
- High-spin single-particle Routhians, alignments, quadrupole moments etc., ignoring the pairing force.
- Total energies, total spin, etc. for rapidly rotating nuclei, allowing the calculation of the nuclear shape as a function of spin.
- Static properties including pairing; i.e., ground-state shapes, pairing gaps, nuclear masses, etc.

The quadrupole (ε and γ) and hexadecapole (ε_4) deformation degrees of freedom may be studied. As the program published here is a multipurpose one, it is not really optimized for doing large-scale calculations of any of the quantities mentioned. Note also that pairing is not included for nuclear spin $I > 0$, which means that, for example, band-crossings caused by pairing cannot be described.

In Sect. 3.2 the formalism of the cranked Nilsson–Strutinsky model is presented; the corresponding computer code, NICRA, is introduced in Sect. 3.3. A few specific points concerning the installation of the code are discussed in Sect. 3.4, two examples of how to run the code are presented in Sect. 3.5, and some further developments are indicated in Sect. 3.6.

3.2   Present Version of the Nilsson–Strutinsky Approach

In the Nilsson–Strutinsky approach, the total nuclear energy is split into an average part, parametrized by a macroscopic expression, and a fluctuating part extracted from the variation of the level density around the Fermi surface. The microscopic part includes the Strutinsky shell correction and the pairing energy. Rotation is introduced in the cranking approximation [3.29], corresponding to rotation around one principal axis.

3.2.1   The Single-Particle Potential

The basic developments leading to the modified single-particle oscillator potential are described in [3.3,6,30], while cranking was introduced in [3.12,15]. The single-particle potential used here is in the form [3.31]

$$h^\omega = h^0 - \omega j_x = h_{\text{ho}}(\varepsilon, \gamma) + 2\hbar\omega_0\rho^2\sqrt{\frac{4\pi}{9}}\varepsilon_4 V_4(\gamma) + V' - \omega j_x \,, \qquad (3.1)$$

where $h_{\text{ho}}(\varepsilon, \gamma)$ is the anisotropic harmonic oscillator potential

$$h_{\text{ho}}(\varepsilon, \gamma) = p^2/2m + \frac{1}{2}m\left\{\omega_x^2 x^2 + \omega_y^2 y^2 + \omega_z^2 z^2\right\} \qquad (3.2)$$

with the frequencies ω_x, ω_y and ω_z expressed in the quadrupole deformation parameters in the usual way with signs chosen according to the Lund convention [3.10,15,32]

$$\omega_j = \omega_0(\varepsilon, \gamma) \left[1 - \frac{2}{3} \varepsilon \cos \left(\gamma + \frac{2\pi \nu_j}{3} \right) \right] , \quad j \in \{x, y, z\} \tag{3.3}$$

with $\nu_x = 1$, $\nu_y = -1$, and $\nu_z = 0$. The calculations are carried out in the stretched coordinate system [3.3,10], $\xi = x\sqrt{M\omega_x/\hbar}$ etc., and the higher multipoles in the potential are also defined in these coordinates; i.e., the spherical harmonics $Y_{\lambda\mu}$ are functions of the angles θ_t and φ_t where the index t refers to the stretched system. The corresponding radius coordinate is denoted by ρ in (3.1).

The hexadecapole potential is defined to obtain a smooth variation [3.33] in the γ-plane so that the axial symmetry is not broken for $\gamma= -120°, -60°, 0°$, and $60°$. It is of the form [3.34,35]

$$V_4 = a_{40}Y_4^0 + a_{42}(Y_4^2 + Y_4^{-2}) + a_{44}(Y_4^4 + Y_4^{-4}) \tag{3.4}$$

where the a_{4i} parameters are chosen as

$$a_{40} = \frac{1}{6}(5\cos^2\gamma + 1) , \qquad a_{42} = -\frac{1}{12}\sqrt{30}\sin 2\gamma ,$$

$$a_{44} = \frac{1}{12}\sqrt{70}\sin^2\gamma . \tag{3.5}$$

It is straightforward to introduce deformations of higher multipoles like ε_6, or odd multipoles like ε_3 or ε_5. The present computer program, however, is limited to the form of (3.1). The term V', which is also defined in the stretched coordinates, is introduced to reproduce the level ordering as observed in nuclei,

$$V' = -\kappa(N)\hbar \overset{\circ}{\omega}_0 \left\{ 2\boldsymbol{\ell}_t \cdot \boldsymbol{s} + \mu(N)(\ell_t^2 - <\ell_t^2>_N) \right\} . \tag{3.6}$$

The parameters κ and μ might either be given the same values for each shell or, alternatively, as indicated in (3.6), they can be made dependent on the main oscillator quantum number $N = N_t$.

The diagonalization of the Hamiltonian (3.1) gives the eigenvalues e_i^ω and the eigenvectors χ_i^ω. Furthermore, the single-particle energies in the laboratory system and the single-particle spin contributions m_i are obtained as

$$e_i = \langle \chi_i^\omega | h^0 | \chi_i^\omega \rangle , \qquad m_i = \langle \chi_i^\omega | j_x | \chi_i^\omega \rangle , \tag{3.7}$$

where h^0 is the static single-particle Hamiltonian. (See (3.1)).

3.2.2 Total Nuclear Quantities

We define the total quantities

$$E_{\text{sp}} = \sum_{\text{occ}} e_i = \sum_{\text{occ}} e_i^\omega + \hbar\omega \sum_{\text{occ}} m_i , \tag{3.8}$$

$$I = \sum_{\text{occ}} m_i \, , \tag{3.9}$$

with the summation over the occupied orbitals in a specific configuration of the nucleus. The shell energy is now calculated from

$$E_{\text{shell}}(I) = E_{\text{sp}}(I) - <E_{\text{sp}}(I)> \, , \tag{3.10}$$

where $<E_{\text{sp}}(I)>$ is the smoothed single-particle sum evaluated according to the Strutinsky prescription [3.4,5]. The detailed formulas for $<E_{\text{sp}}(I)>$ are discussed in Ref. [3.6] for $I = 0$ and in Ref. [3.15,22] for $I \neq 0$.

The pairing energy is an important correction that should decrease with increasing spin and become essentially unimportant at very high spins. To obtain an $(I = 0)$ average pairing gap Δ, which varies as $A^{-1/2}$, the pairing strength G is chosen as [3.6,25]

$$G_{\text{p,n}} = \frac{1}{A} \left(g_0 \pm g_1 \frac{N - Z}{A} \right) \quad (\text{MeV}) \tag{3.11}$$

with $g_1/g_0 \approx 1/3$. Furthermore, the number of orbitals included in the pairing calculation should vary as $\sqrt{Z}$ and $\sqrt{N}$ for protons (p) and neutrons (n), respectively.

The total nuclear energy is now calculated by replacing the smoothed single-particle sum by the rotating-liquid-drop energy and adding the pairing correction,

$$E_{\text{tot}}(\bar{\varepsilon}, I) = E_{\text{shell}}(\bar{\varepsilon}, I) + E_{\text{RLD}}(\bar{\varepsilon}, I) + E_{\text{pair}}(\bar{\varepsilon}, I) \, , \tag{3.12}$$

where $\bar{\varepsilon} = (\varepsilon, \gamma, \varepsilon_4)$. The shell and pairing energies are evaluated separately for protons and neutrons at $I = 0$, while the renormalization of the moment of inertia introduces a coupling when evaluating E_{shell} for $I > 0$. In the present computer program, E_{pair} is included only for $I = 0$. The protons and neutrons are also coupled through the requirement that the shape of the respective potentials and the rotational frequencies are identical.

In the liquid drop model [3.36], the nuclear mass is given by

$$\begin{aligned}
E_{\text{L.D.}} = {}& -a_v \left(1 - \kappa_v \left(\frac{N - Z}{A} \right)^2 \right) A \\
& + \frac{3}{5} \frac{e^2 Z^2}{R_c} \left[B_c(\bar{\varepsilon}) - \frac{5\pi^2}{6} \left(\frac{d}{R_c} \right)^2 \right] \\
& + a_s \left(1 - \kappa_s \left(\frac{N - Z}{A} \right)^2 \right) A^{2/3} B_s(\bar{\varepsilon}) \\
& + \begin{cases} +12/\sqrt{A} & \text{odd–odd nuclei} \\ 0 & \text{odd–even nuclei} \\ -12/\sqrt{A} & \text{even–even nuclei.} \end{cases}
\end{aligned} \tag{3.13}$$

In this formula, $B_c(\bar\varepsilon) = E_{\text{Coul}}(\bar\varepsilon)/E_{\text{Coul}}(\bar\varepsilon = 0)$ and $B_s(\bar\varepsilon) = E_{\text{surf}}(\bar\varepsilon)/E_{\text{surf}}(\bar\varepsilon = 0)$, are the surface and Coulomb energies of a nucleus with a sharp surface in units of their corresponding values for spherical shape. The second term in the Coulomb energy is a (shape-independent) diffuseness correction with d being the diffuseness. The Coulomb energy constant is often defined as $a_c = (3/5)(e^2/R_c)$. When calculating the nuclear mass, one should note that the average pairing energy should be subtracted from E_{pair}. With our default parameters, (3.18) below, it is roughly equal to 2.3 MeV (see Ref. [3.6]).

In the computer program we only consider the *correction* to the $I = 0$ spherical-liquid-drop energy due to rotation and deformation, δE_{RLD}. It reads

$$\delta E_{\text{RLD}}(\bar\varepsilon, I) = a_c \frac{Z^2}{A^{1/3}}(B_c(\bar\varepsilon) - 1)$$

$$+a_s\left(1 - \kappa_s\left(\frac{N-Z}{A}\right)^2\right)A^{2/3}(B_s(\bar\varepsilon) - 1) + \frac{\hbar^2}{2A^{5/3}\mathcal{J}_{\text{rig}}(\bar\varepsilon)}I^2 \qquad (3.14)$$

where $(B_c(\bar\varepsilon) - 1)$ and $(B_s(\bar\varepsilon) - 1)$ are calculated by numerical integration. For $\varepsilon_4 = 0$, they are tabulated in Ref. [3.10].

The calculation of the Coulomb correction in particular is somewhat involved: the original six-dimensional integral can be simplified only to four dimensions [3.37,38]. Furthermore, the use of stretched coordinates leads to complicated expressions.

The radius expressed as a function of the angles in the stretched-coordinate system is obtained by requiring a constant value for the potential in (3.1) (neglecting the $V' - \omega j_x$ term):

$$\rho^2 \propto \frac{1}{1 - \frac{2}{3}\varepsilon\sqrt{\frac{4\pi}{5}}\left(\cos\gamma Y_{20} - \frac{1}{\sqrt{2}}\sin\gamma(Y_{22} + Y_{2-2})\right) + \varepsilon_4\sqrt{\frac{4\pi}{9}}V_4(\gamma)}, \qquad (3.15)$$

where the spherical harmonics are functions of the angles θ_t and φ_t and the harmonic-oscillator part of the potential is expressed in ε and γ. From the definition of the stretched coordinates, it is straightforward to express the angles θ_t and φ_t in the corresponding angles in the spherical system, θ and φ. These expressions, as well as the final formula for the radius, are given in the appendix of Ref. [3.11]. In the formula given there, only hexadecapole terms of Y_{40}-type are included. However, as the stretching is influenced only by the quadrupole coordinates ε and γ, one can simply replace the expression "$2\varepsilon_4 P_4(\cos\theta_t)$" of Ref. [3.11] with that used here (3.1,4,5) (cf. (3.15)). Similarly, other multipoles are easily included in the expression for the radius by simply adding, in a way analogous to that for the hexadecapole term, additional terms in the single-particle potential.

Because of the incompressibility of nuclear matter, the nuclear volume is kept constant when the nucleus is deformed. This is achieved by varying the frequency $\omega_0(\varepsilon, \gamma, \varepsilon_4)$ from its value for a spherical shape, $\overset{o}{\omega}_0$. The integration of the nuclear volume is most easily performed in the stretched-coordinate system and then multiplied with the corresponding Jacobian, a constant proportional to $\sqrt{\omega_x \omega_y \omega_z / \omega_0^3}$.

From the single-particle wavefunctions the electric (or mass) quadrupole moment may be calculated as

$$Q_2 = \sum_{\text{occ}} p_i \langle \chi_i^\omega | q_2 | \chi_i^\omega \rangle, \tag{3.16}$$

where $p_i = 1$ for protons and 0 (or 1) for neutrons. In the code we have chosen $q_2(\hat{x}) = \sqrt{(3/2)}(y^2 - z^2)$, where x is the rotation axis. At sufficiently high spins we may then get an estimate for the in-band $\Delta I = 2$, $E2$-transition probability [3.58]: $B(E2; I \to I - 2) \approx (5/16\pi) |Q_2(\hat{x})|^2$ where $Q_2(\hat{x})$ is the electric quadrupole moment. This approximation is expected to be valid for high spins and substantial triaxial deformations [3.59]. At $\gamma = 0°$ the "usual" quadrupole moment, $q_0(\hat{z}) = 2z^2 - x^2 - y^2$, is obtained as $Q_0(\hat{z}) = -\sqrt{(8/3)}Q_2(\hat{x})$. For triaxial shape, $Q_0(\hat{x})$ may be calculated *instead* of $Q_2(\hat{x})$ by a simple change in the subroutine SETMAT_NILS (see comments in this routine). At high spin and at large triaxial deformation, $Q_0(\hat{x})$ is equal to the static ($\Delta I = 0$) quadrupole moment seen in the laboratory frame.

3.2.3 Parameters

The parameters appearing in the calculations can be divided into three fundamentally different types:

1. Parameters like κ and μ describing the $\boldsymbol{\ell} \cdot \boldsymbol{s}$ and ℓ^2-strengths of the single-particle potential or the liquid-drop surface-energy constant a_s. Such parameters are fixed from fits to experimental quantities and are valid over large ranges of nuclei.

2. Parameters like those describing the deformation of the mean field. Such parameters may be calculated from self-consistency conditions (e.g., minimization of the total energy), or they may be fixed, for example, from measured quadrupole moments. On some occasions, within reasonable limits, one might also use them as fitting parameters.

3. Parameters like the Strutinsky-smearing width parameter γ_s or the number of N-shells used in the single-particle diagonalization. These are auxiliary parameters that should be given values so that convergence is obtained. For example, a sufficient number of N-shells should be included so that the orbitals around the Fermi surface are insensitive to the addition of more shells or, if the total energy is calculated, the shell and the pairing energies have converged in the same sense.

In the present section, we mainly discuss parameters of type 1. Let us start with those of the single-particle potential. The default values of $\kappa(N)$

and $\mu(N)$ in the program are given in Table 3.1 below. These parameters are used, for example, in Ref. [3.31] and in many subsequent papers. For very light nuclei, κ is larger than used in many previous publications, e.g. [3.11,39]. This makes the spherical magic gaps at particle numbers 8 and 20 smaller and tends to improve, for example, the mass fits [3.23]. For heavy nuclei, we have been guided by the fits in Ref. [3.6]. It has been known for some time that the present parameters are not optimal for the actinide region (see Ref. [3.20] for a detailed discussion). In this region, the parameters suggested by Rozmej [3.40], $\kappa_p = 0.0580, \mu_p = 0.630, \kappa_n = 0.0526, \mu_n = 0.457$, seem to give a better fit. These values have been used for all shells, but another possibility is to keep a similar variation as in the example input presented below in Table 3.1, changing κ and μ only for the shells relevant in the actinides. Discussions of other alternatives of the κ and μ parameters in different mass regions can be found, for example, in [3.41–45].

The only additional parameter for the single-particle potential is the oscillator strength $\overset{o}{\omega}_0$. By default it varies as

$$\hbar \overset{o}{\omega}_{0p/n} = \frac{41}{A^{1/3}} \left(1 \mp \frac{N-Z}{3A} \right) \quad (\text{MeV}) \tag{3.17}$$

for protons and neutrons, respectively. This gives an approximate mean-square radius of $1.2 A^{(1/3)}$ fm. The larger value for neutrons than protons simulates the Coulomb potential by giving a shallower potential well for the protons.

If pairing is calculated, the pairing strength G is an important parameter. Its value is given according to (3.11). As a default we use

$$g_0 = 19.2, \quad g_1 = 7.4 \tag{3.18}$$

with $\sqrt{15Z}$ and $\sqrt{15N}$ orbitals considered below and above the Fermi surface. These numbers are chosen to give an average pairing gap of approximately $12/\sqrt{A}$ MeV. If this recipe is used for Z or N smaller than 60, there are not enough levels below the Fermi surface; in this case, all levels below the Fermi surface, and an equal number above, are included in the pairing calculations. This yields an average pairing gap that decreases slowly (relative to $12/\sqrt{A}$ MeV) with decreasing mass, a feature which seems reasonable (see [3.23]). An alternative recipe is to use

$$g_0 = 22.0, \quad g_1 = 8.0 \tag{3.19}$$

with $\sqrt{15Z}$ ($\sqrt{15N}$) replaced by $\sqrt{5Z}$ ($\sqrt{5N}$). This produces about the same average gap as the formula above, but acting with full strength for Z or $N \geq 20$. A different and probably somewhat more consistent recipe to obtain a prescribed average gap is used in Ref. [3.8]. One can also note recent fits of the pairing gap suggesting that it should vary not only with mass number but also with isospin [3.46,47]. In the present computer program, however, only the method with input of G and corresponding range is implemented.

The liquid-drop parameters used are those of Myers and Swiatecki [3.36]:

$$a_v = 15.4941 \quad \text{MeV}, \quad a_s = 17.9439 \quad \text{MeV},$$
$$\kappa_v = \kappa_s = 1.7826, \qquad d = 0.546 \quad \text{fm},$$
$$R_c = 1.2249 A^{1/3} \quad \text{fm} \quad (a_c = 0.70531 \quad \text{MeV}).$$

It is our experience that the simple formulae (3.13,14), give a very reasonable fit over the entire nuclear chart. For a more detailed description of, for example, the fission process, elaborate formulae should be used (cf. Ref. [3.48] and refs. therein). However, the present code is not intended for such applications and we have seen no reason to include possibilities other than the usual liquid-drop expression. The rigid-body moment of inertia is calculated with a radius parameter, $r_0 = 1.2$ fm, and with a sharp surface.

We now turn to the parameters of type 3. With stretched coordinates, the quadrupole part of the potential is already treated exactly within one oscillator shell, $N = N_t$. It is then only the hexadecapole potential and the cranking term that couple the N-shells, $\Delta N = \pm 2$. The wave function and energy of a single-particle orbital are then described with very high accuracy if, for an orbital mainly belonging to the N' shell, $(N'+2)$ shells are considered in the diagonalization. For the pairing calculations, it is important to have so many shells so that no orbital within the "pairing range" is omitted. However, it is not really necessary to include also the $\Delta N = 2$ partners of all orbitals in the range. We also note that the stretched coordinates have the disadvantage that the basis functions vary with deformation, ε and γ.

In the Strutinsky renormalization, the smoothed single-particle sum (3.10) is sensitive to discrete orbitals in a range around the Fermi surface proportional to the smearing parameter γ_s [3.4–6]. By default, γ_s is given the value $1.0\,\hbar\omega_0$ with a sixth-order smearing polynomial. Even though complete convergence requires accurate single-particle orbitals rather high above the Fermi surface, it turns out that, for the modified oscillator potential, the shell energy can be calculated with a reasonable accuracy using a rather small basis. Some idea of the uncertainties can be obtained from Fig. 3.1, where the shell energy is drawn as a function of deformation for different numbers of N-shells in the diagonalization. As an alternative to $\gamma_s = 1.0\hbar\omega_0$, the slightly larger value $\gamma_s = 1.2\,\hbar\omega_0$, has often been used (see [3.6]). Furthermore, for very light nuclei it was found in Ref. [3.11] that the even larger value $\gamma_s = 1.6\,\hbar\omega_0$ gave a more reliable shell energy. This will increase the sensitivity to higher shells but, because of the small values of Z and N, will not require very large values of $N_{\max}$.

3.3   The Code NICRA

The present version of the *Nilsson Cranker* computer code has been developed from a long tradition of Lund computer codes initiated by S. G. Nilsson. The

Table 3.1. The structure of the input file for **NICRA** with the case described in Sect. 3.5.1 as a specific example. Note that because IN_LEV = 2, lines 12–25 will not be read, but they are still included to show the full possibilities. The values listed for κ and μ are those used as default in the code. In the short explanation of the input to the right in the table (following the asterisks), default values are indicated by D.

```
1,   70,90,21,2,        *Z,N,LEV_PRINT,IN_LEV (D: , ,21,0)
2,   1,0.20,            *NEPS,(EPSI(I),I=1,NEPS)
3,   1,0.0,             *NGAM,(GAMI(I),I=1,NGAM)
4,   1,0.0,             *NEPS4,(EPS4I(I),I=1,NEPS4)
5,   -3,0.0,0.7,0.1,    *NOMEGA,(OMEGA(I),I=1,NOMEGA)
0,,,,,,,,               *IN_LEV=0 stops here
6,   2,3,0.             *IDEF_TYP,NVAR,FI (D:0,3,0.)
7,   'nicra.spag',1,3., *filename,IF_SPOUT,SPERANGE (D:' ',0,3)
8,   'nicra160yb', 0,   *filename,IF_LEVSAV (D:' ',0)
9,   'nicra.ener',1,10, *filename,IF_TOTEOUT,ISTEP (D:' ',0,0)
0,,,,,,,,,,,,,,         *IN_LEV=1 stops here
10,  1,2,1,1            *IF_MEV,IF_STRUT,IF_Q2,IF_BCS (D:1,2,1,1)
11,  1,2,2,1            *IF_DROPCALC,ISO1,ISO2,IF_ISOENEROUT (D:1,1,2,0)
1000, 1,,,,,,,          *IF_GO,(0=>stop for testing: only read and write)
0,,,,,,,,,,,,,,,        *IN_LEV=2 stops here
12,  4, 8, 8,,,,,,      *NUU,NMAX(pro),NMAX(neu),G0,G1,KOEFF_RANGE
0,,,,,,,,,,,            *(D:4,MAXN,MAXN,19.2,7.4,15)
13,  0.120,0.00, 0.120,0.00 *(kappa(N=0,t),my(N=0,t),t=pro,neu)
14,  0.120,0.00, 0.120,0.00 *   kappa(N=1,pro),my...
15,  0.105,0.00, 0.105,0.00 *   .
16,  0.090,0.30, 0.090,0.25 *   .
17,  0.065,0.57, 0.070,0.39 *   .
18,  0.060,0.65, 0.062,0.43 *   .
19,  0.054,0.69, 0.062,0.34 *   .
20,  0.054,0.69, 0.062,0.26 *   .
21,  0.054,0.69, 0.062,0.26 *   .
22,  0.054,0.69, 0.062,0.26 *   .
23,  0.054,0.69, 0.062,0.26 *   .
24,  0.054,0.69, 0.062,0.26 *   kappa(N=11,pro),my...
25,  0.054,0.69, 0.062,0.26 *   kappa(N=12,pro),my...
```

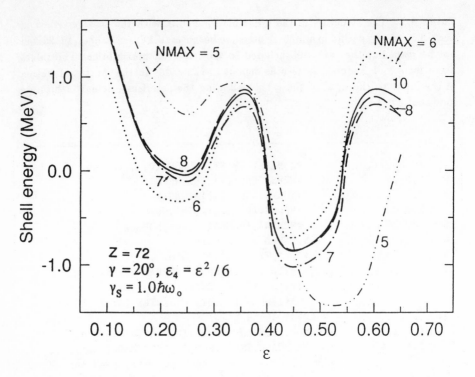

Fig. 3.1. The proton shell energy ($\omega = 0$) for $Z = 72$ drawn as a function of ε with $\gamma = 20°$ and $\varepsilon_4 = \varepsilon^2/6$. Different numbers of oscillator shells are used, NMAX = 5, 6, 7, 8, and 10. The first $N=6$ orbital becomes occupied at $\varepsilon \approx 0.4$ so the NMAX = 5 case is included only as a warning for inexperienced users. Note, however, that NMAX = 6 already gives reasonable results, although it is of course safer to use at least NMAX = 8. The example is chosen to illustrate the shell-energy pocket for Z=72 at a large triaxial deformation [3.49] which becomes even more pronounced at high spins [3.50].

single-particle Hamiltonian, h^ω (3.1) is diagonalized in a stretched harmonic-oscillator basis $|N_t\,\ell_t\,j_t\,\Omega_t\rangle$ for a given deformation (ε, γ, ε_4) and rotational frequency (ω). This leads to single-particle quantities like energy (e_i (3.7)), angular-momentum component along the axis of rotation (m_i (3.7)), signature (α_i), parity (π_i), and quadrupole moment (q_i) calculated separately for protons and for neutrons. Then for the specified nucleus (Z and N) the single-particle quantities are summed giving $E_{\rm sp}$, I, α, π and Q_2, and the shell energy $E_{\rm shell}$ is calculated through the Strutinsky procedure. In addition the rotating-liquid-drop energy is calculated and optionally, if $\omega = 0$, also the pairing energy. The sum of these three quantities gives the total energy $E_{\rm tot}(\varepsilon, \gamma, \varepsilon_4, I, \alpha, \pi)$ (3.12). One may be interested in studying either total or single-particle quantities. In both cases the above-mentioned calculation may be performed on some kind of lattice; i.e., in a loop over one or several of the variables $\varepsilon, \gamma, \varepsilon_4$, and ω. In high-spin calculations we usually want to loop at least in ω. For a given deformation, the parameters in the rotating-liquid-drop energy expression (3.14), as well as the volume conservation factor $\omega_0/\overset{o}{\omega}_0$, only have to be calculated once.

3.3.1   Structure of the Code

The structure of the code NICRA is described schematically in Fig. 3.2 in terms of a flow chart showing the connections between the major subroutines and the main program. The deformation and frequencies of a specific run are read in LOOP_READER and the κ- and μ-values (which may also be set to default values) are read in NILS_READER. After some service tasks the deformation loop starts by setting deformations (in LOOP1_SET) according to input specification. The smooth variation in the γ-plane of the Y_{4m}-components ($m = 0$, 2, and 4 (3.4,5) is treated in DEFSET. Volume-conservation routine OMO yields ''OMOM'' ($\omega_0/\overset{o}{\omega}_o$) and the deformed liquid-drop parameters (3.14) ''BS'' ($B_s - 1$), ''BC'' ($B_c - 1$), and ''XJ'' ($\mathcal{J}_{\rm rig}(\bar{\varepsilon})/\mathcal{J}_{\rm rig}(\bar{\varepsilon} = 0)$) are calculated in DROP.

In the isospin loop, which performs calculations first for protons and then for neutrons, two important subroutines are called: SETMAT_NILS and ORBITS. In the former, matrix elements of the terms h^0 and j_x of the single-particle Hamiltonian ((3.1)) are set up for a cranked-Nilsson-model calculation and stored in the fields EIJ and RJXIJ, respectively. If the quadrupole moment is to be calculated, a matrix for this operator is also set up here and stored in Q2IJ. Since only even multipolarities of spherical harmonics are allowed in the NICRA code, parity and signature are always good quantum numbers. Even N_t-values correspond to positive parity and odd N_t-values to negative parity. To facilitate the calculation of matrix elements we let the rotation axis and the quantization axis coincide. If the x-axis is the rotation axis, this means that $\Omega = m_i$ is now the angular-momentum component along the x-axis. Thus, for a prolate nucleus cranked around the x-axis (perpendicular to the symmetry axis, denoted as the z-axis), the Ω-values ($= \Omega_x$) used in the program have a different meaning than the good quantum numbers at spin zero $\Omega = \Omega_z$. In the program this is solved by working internally with $(\gamma + 120)°$ as the triaxiality parameter, which corresponds to a cyclic permutation of the intrinsic axes. The signature quantum number $\alpha = 1/2$ is simply associated with $\Omega = 1/2, 5/2, 9/2 \ldots$ while $\alpha = -1/2$ corresponds to $\Omega = 3/2, 7/2, 11/2 \ldots$. (If the user wants to add some terms to the present Nilsson Hamiltonian, they are simply added in the routine SETMAT_NILS. If the added terms break a symmetry, the looping and sizes of fields also need to be changed. If the deformation coordinates are changed, changes are required also in the routines OMO and DROP.)

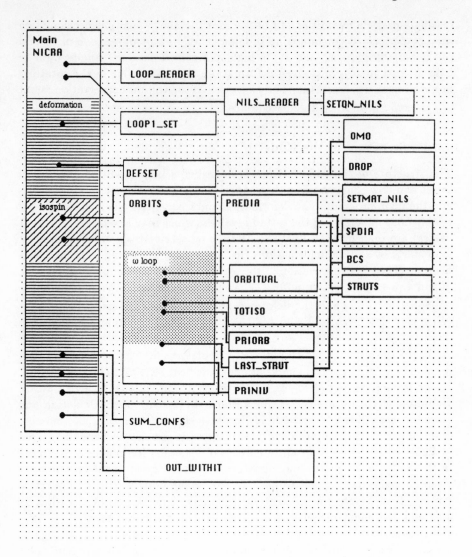

Fig. 3.2. The general structure of the NICRA code as described in more detail in Sect. 3.3.1.

The subroutine ORBITS is the main driver for an ω-loop (for one kind of particle). For each ω-value the matrices $h^0 - \omega j_x$ are diagonalized in SPDIA for each set of good quantum numbers (i.e. π and α). Expectation values are calculated for the angular-momentum component along the x-axis, $\Omega_x = m_i = \langle j_x \rangle$, and quadrupole moment along the x-axis, $q_2(\hat{x})$ (in subroutine ORBITVAL), and summed quantities of E, I and Q_2 for one kind of particle are calculated in TOTISO. Optionally, single-particle quantities of e, j_x, and q (for each symmetry) are printed in PRIORB. A Strutinsky smearing of energies and spins in subroutine STRUTS is normally performed, called by PREDIA for $\omega = 0$ when IF_STRUT=1, and, for IF_STRUT=2, also called by LAST_STRUT at the highest ω-values. In the latter case, a linear fit is made, (cf. (3.10)),

$$\langle E_{\rm sp}(\tilde{I}) \rangle = A + B\tilde{I}^2 , \qquad (3.20)$$

where $\tilde{I}$ is the smoothed spin. Here, $A = \langle E_{\rm sp}(I = 0) \rangle$ while B is the smooth moment of inertia factor, $B = \hbar^2/2\mathcal{J}_{\rm STRUT}$. Equation (3.19) could be regarded as the first terms in a series where higher terms are essentially negligible (for very light nuclei, higher-order terms must be included). However, these terms mean that B will be slightly dependent on the highest ω-value chosen. Owing to the $\bar{\ell}^2$-term, the smooth moment of inertia is usually about 30% larger than the rigid-body value [3.15]. In a "normal" run both of these quantities are printed for each deformation. At spin zero, a BCS pairing calculation is done in the subroutine BCS (unless IF_BCS is changed from default value 1 to 0).

After ORBITS, total quantities are calculated in SUM_CONFS. From the summed energies and spins, $E_{\rm sp}$ and I (3.8,9) the total energy (3.12) is calculated as

$$E_{\rm tot}\,(\varepsilon, \gamma, \varepsilon_4, I) = E_{\rm sp} - A + E_{\rm L.D.} - BI^2 + \frac{\hbar^2}{2\mathcal{J}_{\rm rig}}I^2 \qquad (3.21)$$

Printing is performed in PRINIV for proton and neutron quantities and in OUT_WITHIT for total quantities.

3.3.2   Input

An example of an input file, together with short explanations of each input line, is given in Table 3.1. This input corresponds to the first example in Sect. 3.5 below, leading to Fig. 3.3. For error detection and convenience, each input line starts with a running integer. On the first line the nucleus is specified by proton number and neutron number. Then follow two other integers, LEV_PRINT and IN_LEV, which determine the amount of output and input, respectively; the larger the number the more output/input. For a "normal" run LEV_PRINT is either 20 or 21. If it is set to 20, only total quantities (energy, spin, etc.) are printed while single-particle energies and spins are also printed if LEV_PRINT = 21. With increasing values of LEV_PRINT, the

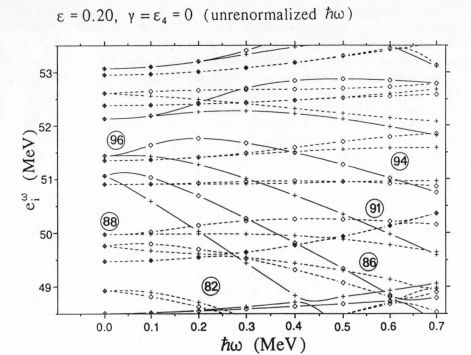

$\varepsilon = 0.20, \quad \gamma = \varepsilon_4 = 0$ (unrenormalized $\hbar\omega$)

Fig. 3.3. Single-neutron orbitals drawn as functions of rotational frequency for $\varepsilon = 0.20$, $\gamma = 0°$, $\varepsilon_4 = 0$. Positive (negative) parity states are drawn by *dashed (full) lines* and *plus (open diamond) signs* indicate signature $\alpha = +1/2$ ($\alpha = -1/2$). For more details, see Sect. 3.5.1 and Table 3.1.

amount of output increases, while output from a specific subroutine may also be obtained by setting LEV_PRINT to the number of that subroutine (specified in the first lines of each subroutine). In the file "APPEN", an output example is given corresponding to LEV_PRINT = 21. IN_LEV may have the values 0, 1, 2, or 3. Generally, a larger value means that more parameters may be changed from their default values and that more details about the run may be given. For the lowest level, only deformation values are given as input; one line each for $\varepsilon, \gamma, \varepsilon_4$, and ω-values. These values may be explicitly specified starting with the total number of ε-values (NEPS), etc. Alternatively, by starting one of the lines with "−3" the specific deformation will run from the lowest (START) to the highest (STOP) value in steps of STEP (see line 5 in Table 3.1). In a run with IN_LEV=0 all quantities will be calculated for the specified deformations and frequencies but nothing will be stored for further processing.

After the "final" input line, a line starting with 1000 is expected to close the input. Thus, for IN_LEV = 0, this should be the sixth line. Next level (IN_LEV=1) allows different ways of arranging the loops (line 6) and further processing in several ways, with up to 5 files being used for writing. These written files are specified on lines 7, 8, and 9, and described in the next section.

On line 6, IDEF_TYP makes it possible to arrange the loops over deformations and frequencies in different ways (see the description in the beginning of the main program). By default it is set equal to 0, corresponding to independent loops over ε, γ, ε_4, and ω. For IDEF_TYP = 1, ε and γ vary in the same loop (NEPS = NGAM) with an independent loop over ε_4 inside. The parameter NVAR defines a main variable. It is useful, for example, if one wants to plot single-particle orbitals as functions of this variable. The most important case, however, is NVAR = 3 (default value) with ω as the main variable. Then it is possible to calculate E_{tot} vs. I, do a Strutinsky renormalization of the moment of inertia, etc. The variable FI is used together with IDEF_TYP = −1, in which case a rectangular mesh, tilted by the angle FI, is created in the (ε, γ)-plane. The angle is defined between the $\gamma = 0$ axis and the x-axis, e.g. $\gamma = 60°$ coincides with the ordinate for FI = −30°.

The default values may be further changed by putting IN_LEV = 2. No Strutinsky renormalization of the moment of inertia is performed for IF_STRUT = 1 ($B = \hbar^2/2\mathcal{J}_{rig}$ in (3.20)) while also the spin-zero renormalization is switched off ($A = E_{L.D.}$ in (3.20)) for IF_STRUT = 0. No quadrupole moments, no pairing energies or no liquid-drop parameters are calculated if IF_Q2, IF_BCS, or IF_DROPCALC, respectively, are set to zero. If IF_MEV=0, single-particle energies are calculated in units of $\hbar\omega_0$ and single-particle quadrupole moments in units of $\hbar/m\omega_0$. In this case, no total quantities are calculated. Only protons (neutrons) are considered when ISO1 = ISO2 = 1 (2) while output of proton and neutron quantities on file occurs for IF_ISOENEROUT = 1 (if LEV_PRINT ≥ 21).

The last level of input (IN_LEV=3) allows for changes of the number of N_t-shells that are coupled (NUU) and the highest included N_t-shell for protons or neutrons (NMAX). The parameters G0, G1 and KOEFF_RANGE determine the pairing strength (see (3.11)) with the default values given by (3.18).

The next NMAX+1 lines give Nilsson-model parameters $\kappa_p, \mu_p, \kappa_n, \mu_n$ for each N_t-shell starting with $N_t = 0$.

In the last line in each case having 1000 as the running number, one should put IF_GO=1. For IF_GO=0, input is only read and written but no calculation is done (useful for prechecking long runs).

3.3.3   Files Used

In the input lines 7, 8, and 9, the first parameter gives the name of a file that is only opened if the second parameter is not equal to 0. In the first file (line 7) single-particle energies and spins are written in the energy interval $\pm$ SPERANGE (MeV) around the Fermi surface. This file may conveniently be used for construction of various types of Nilsson diagram; i.e., e_i^ω (or e_i) vs. $\varepsilon, \gamma, \varepsilon_4$, or ω. In this file, after the first line of information (A-format) several lines come, each corresponding to one curve. The use of this file is exemplified in Fig. 3.3 below, showing e_i^ω vs. ω, for neutrons at $\varepsilon = 0.20$, $\varepsilon_4 = 0$, and $\gamma = 0°$. Each line starts with a quantum number, 100 for $(\pi, \alpha) = (+, 1/2)$, 200 for $(-, 1/2)$, ..., and is followed by (e_i^ω, m_i)-values at the calculated deformation points. One may thus either plot e_i^ω-values (with or without using the m_i-values, which are the negative of the slopes), or the m_i-values. The input for a given plot is ended by a line starting with 0. Note that if one wants to study e_i^ω vs. ω the frequencies are written unnormalized. Owing to the presence of the ℓ^2-term in the Nilsson model, the moment of inertia (and thus the rotational frequency) must be renormalized. All frequency-values written in the file should thus be multiplied by a factor written on output. The factor is around 1.3, and can be found in the printout as $\mathcal{J}_{\text{str}}/\mathcal{J}_{\text{L.D.}}$ (see (25,26) in Ref. [3.51]).

The second output file is most important when larger calculations are to be performed. If IF_LEVSAV=1, various information from the run is saved in files and may easily (and more quickly) be used for another nucleus (with identical deformations and frequencies) in another run. In this later run single-particle values are read from files (if IF_LEVSAV=−1) instead of being calculated. For reasons of convenience, and also compatibility with other codes used at Lund, this output is written on three different files given the names EN-FIL, LQ-FIL and FIL if "FIL" was the first variable on line 8. In these three files single-particle quantities, liquid-drop quantities, and information about the various deformation points, respectively, are written. In the final output file (line 9) total quantities are written, which may be used, for example, to construct diagrams like those in Fig. 3.4.

The data written in files given in lines 7 and 9 are in text format (used for plotting and easily edited), while the data in the files specified at line 8 are written in binary form.

3.3.4   Output

As was mentioned in the previous subsection the amount of output is controlled by the input parameter LEV_PRINT. The output is to a large extent self-explanatory, as can be seen in the example shown in the file "APPEN" and described in Sect. 3.5.2. The list with single-particle quantities, however, may need some explanation. For each deformation, $(\varepsilon, \gamma, \varepsilon_4)$, single-particle states within the energy range $\pm$SPERANGE (default value: 3) are listed together with the energy in the rotating frame (e_i^ω) and angular-momentum component along the rotation axis $(m_i$ or $< j_x >)$ for each rotational frequency calculated. The first row shows the number of the single-particle state, counted from the bottom of the potential at lowest frequency. Each state is identified by a 5-figure number with sign, where the sign denotes the parity, the first figure signature ("1" for $\alpha = +1/2$ and "2" for $\alpha = -1/2$), and the remaining 4 figures are an index number within each group of spin-parity states. After the lists of single-particle (and total) quantities for protons and neutrons at each deformation come the "final results", which show the expectation value of I (along the x-axis) for the calculated deformations (also listed) and frequencies. Note, however, that the frequency shown in this table is usually the renormalized frequency, which deviates from the input frequency by about 30%. The total energy, E_{tot}, is the Strutinsky renormalized energy in the laboratory frame if I_STEP ≥ 0, while it is the Routhian energy, $R = E - \omega I$, for I_STEP < 0. The Routhian is the appropriate quantity to be minimized for fixed cranking frequency. The energy increment between two calculated frequencies is given by dE, while $Q2$ is the electric or mass quadrupole moment (if IF_Q2=1 or 2, respectively). The last row provides information about the proton and neutron configurations in terms of the good quantum numbers (i.e., the total parity and signature). The three digits to the left and right of the decimal point are for protons and neutrons, respectively. The first digit in each case denotes twice the signature and the third digit the parity: "0" for $\pi = +$ and "1" for $\pi = -$. For an even–even nucleus the $\omega = 0$ case is thus always denoted as 000.000; i.e., $\pi = +$ and $\alpha = 0$ for both protons and neutrons.

If I_STEP in the input is larger than zero, an interpolation to specified spin values in steps of I_STEP is performed. Note that quantum numbers, as well as details of the calculated configurations, are completely disregarded in the interpolation. This approximation is thus very crude in some cases, particularly for light nuclei, but may serve as a rather good approximation in heavier nuclei, which rotate collectively.

If a better approximation is needed, configurations have to be followed throughout the crossings in the e_i^ω vs. ω diagram, and small interactions be-

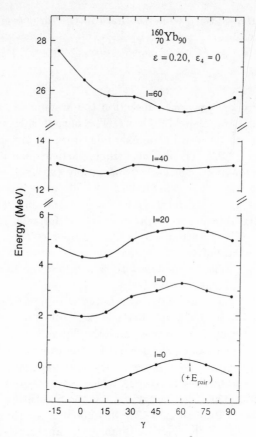

Fig. 3.4. Total energy at different spins for the nucleus $^{160}_{70}\text{Yb}_{90}$ drawn as a function of γ for $\varepsilon = 0.20$, $\varepsilon_4 = 0$. For $I = 0$, the two curves give the energy with and without pairing included. This requires two runs with IF_BCS = 0 and IF_BCS = 1, respectively, or alternatively the printed proton and neutron BCS energies (LEV_PRINT $\geq$ 21) might be added (or subtracted) by hand. For $I > 0$, no pairing is included.

tween single-particle levels have to be removed (the diabatic-level scheme). This program has been carried through [3.31,39,51] but requires some refinements of the present code.

3.4   Installation

Installation of the code should be straightforward on most computer systems. A few specific points should be considered though, as some differences do exist between FORTRAN77 environments. The important ones for the NICRA code are:

- **Variable names:** The six-character restriction in FORTRAN77 is *not* followed and underscores (_) are used to enhance the readability of the code. Truncation of variable names to six characters and/or omission of the underscores does not, however, give rise to any naming conflict.
- **Include files:** To facilitate the declaration of "global" variables in all subroutines where they are needed, these declarations are placed in a separate file, NICRAINC. Furthermore the parameters that control the array dimensions are placed in the file NICRAPAR. These two files are included in the code at compilation time via an INCLUDE command to the compiler. This command is different for different compilers and the INCLUDE lines must therefore be changed accordingly. As a last resort, if the INCLUDE command does not exist, the include files can be copied into the appropriate places in the code.
- **File OPEN statements:** These may vary for different compilers. Special attention should be given to the files opened if IF_LEVSAV $\neq$ 0, as there exist different conventions regarding the unit for the RECL= qualifier in the OPEN statement. In VMS FORTRAN this unit is 32 bit words, whereas standard FORTRAN77 requires bytes (8 bits). This fact is commented in the routine NILS_READER where the files are opened. We have tried to make the two extra filenames so general that they would be accepted on most systems. If hyphens (-) are not accepted in filenames, the creation of filenames must be modified as well.

Both computing times and the size of the executable module strongly depend on the number of main oscillator shells used and declared for, respectively. The size of the module may be controlled easily by changing the parameters MAXN and MAXDIM in the include file NICRAPAR as described in the beginning of the main code. In the code provided, MAXN is set equal to 8 so that MAXDIM=95. For 6 shells declared, the executable module is about 250 kbytes, for 8 shells $\simeq$ 600 kbytes, while for 10 it increases to roughly 1.5 Mbytes. On PC systems, this may restrict the region of nuclei accessible to study. By default, if IN_LEV < 3, the program uses MAXN shells in the calculation, but the number of shells used can be decreased for IN_LEV=3 by giving NMAX(p, n) $\leq$MAXN on line 12 of the input file.

3.5   Examples

Two examples of how to run the code are discussed: how to construct a single-particle diagram e_i^ω vs. ω_{rot} (see Fig. 3.3) and how to minimize the energy with respect to deformation (in this example γ-deformation; see Fig. 3.4) at different spin values. The input data for the first example is shown in Table 3.1 while input and output for the second example are given in file "APPEN".

3.5.1 e_i^ω vs. ω_{rot}

The physics interest in this case is to investigate details of single-particle levels in nuclei with high angular-momentum around ^{160}Yb at a fixed deformation, $\varepsilon = 0.20, \gamma = 0°, \varepsilon_4 = 0$ (lines 2, 3, and 4 in input). The set of frequencies is chosen between 0 and 0.7 MeV in steps of 0.1 (line 5). We choose the input parameters LEV_PRINT and IN_LEV as 21 and 2, respectively, so that single-particle information will be written in the output file and because IF_SPOUT = 1 (line 7) also in the plotting file "nicra.spag". This is a rather fast run and we have no intention of repeating it for a neighboring nucleus. We thus put IF_LEVSAV=0 (line 8). Neither are we interested in plotting total quantities, so IF_TOTEOUT=0 (line 9). As we are only interested in the neutron orbitals, we set ISO1 = ISO2 = 2. On a VAX 3100 workstation the run took 4 minutes (MAXN = 8) (the VAX 3100 is approximately 3 times faster than a VAX 11/780). After the run was completed we used a system-specific plotting routine to obtain the neutron diagram shown in Fig. 3.3. Note how the intruder orbitals originating from $i_{13/2}$ behave with varying frequency (cf. the discussion in Ref. [3.15]).

3.5.2 $E_{\text{tot}}(I)$ vs. γ

In this example we want to investigate how the triaxiality of the nucleus ^{160}Yb changes with spin. We thus choose the input according to the file "APPEN" with LEV_PRINT = 21. This will enable us (for illustrative purposes) to print proton and neutron total quantities if we put IF_ISOENEROUT = 1. We fix the quadrupole deformation to $\varepsilon = 0.2$ and $\varepsilon_4 = 0$ (see e.g. Ref. [3.17]) and vary γ between $-15°$ and $60°$ in steps of $15°$ (line 3). For each γ-value, 9 frequencies between 0 and 0.8 MeV are chosen (line 5), and we want to get interpolated energy values at every 10 units of spin (I_STEP=10). It is reasonable to save the single-particle data in a file to facilitate a run for a neighboring nucleus. We therefore put IF_LEVSAV=1 and expect output in three files EN-nicra160yb, nicra160yb and LQ-nicra160yb. We will produce no single-particle plots, so IF_NIVPRI=0 while IF_TOTEOUT=1 because plots of E_{tot} vs. γ are wanted.

The run took 55 CPU minutes on a VAX 3100 workstation. When it later was repeated for ^{160}Er (IF_LEVSAV = -1), it only took 4 minutes since all information was saved. In Fig. 3.4, E_{tot} vs. γ is shown for ^{160}Yb at $I = 0$ (pairing included), $I = 0, 20, 40$, and 60 (no pairing included). When the curves are constructed, the important symmetry at $\gamma = 60°$ is taken care of by adding two points in the plot: $\gamma = 75°$ and $\gamma = 90°$, which are identical to $\gamma = 45°$ and $30°$, respectively. At $I = 0$ the energy is also symmetric around $\gamma = 0°$ but this symmetry is violated at $I > 0$, since positive γ corresponds to rotation around the smallest axis and negative γ to rotation around the intermediate axis.

Note how the equilibrium deformation smoothly increases from $\gamma = 0°$

at $I = 0$ to $\gamma = 60°$ at $I = 60$, illustrating in a qualitative way the transfer to oblate shape [3.15] and corresponding band terminations [3.17]. Note however that the interpolation becomes rather crude around the particle–hole region ($\gamma = 60°$) and also that if a full minimization in the (ε, γ) plane were performed, ε would decrease as γ increases from $0°$ to $60°$.

In the output shown in file "APPEN", the proton and neutron total quantities are also listed. The quantity E_{tot} is simply the sum of the single-particle energies normalized to the shell energy at $\omega = 0$. If pairing is included, the pairing energy is added for $\omega = 0$ but not for $\omega \neq 0$. "R-shell" is the quasi-shell energy [3.22] or Routhian shell energy, $\sum e_i^\omega - < \sum e_i^\omega >$. This is not the quantity we use in total-energy calculations (cf. (3.20)), but it gives a good idea of the variation of the shell energy with spin. For $\omega = 0$, it coincides with the standard shell energy. In this case, the BCS energy also is listed.

3.6   Further Developments

The NICRA code is intended to serve both as a simple example of a cranked Nilsson–Strutinsky calculation and as a platform for further developments in this field. It is a simplified version of codes used in our recent high-spin calculations (see Refs. [3.44,51] and references therein) and incorporates the most important and basic parts of a Nilsson–Strutinsky calculation. However, some further developments may be needed for a detailed comparison with experimental data. For low spins, pairing is known to be of definite importance and to study excited high-spin bands one needs some kind of configuration restriction in the minimization procedure.

Perhaps the line of development that is easiest to incorporate is to keep track of conserved quantum numbers in the calculations of total quantities. In the present code, where parity and signature are conserved (printed for the yrast configuration in the final output), this would then yield 16 yrast lines instead of one, one for each combination of parity and signature for both protons and neutrons. It is mainly a book-keeping task to introduce this facility. One could also consider studying only the four yrast lines given by the total parity and signature. The physical importance behind this development is to avoid the mixing of states with different symmetry properties in, for example, a shape minimization. This may have a pronounced effect on the result especially for light nuclei or close to the noncollective axes ($\gamma = -120°$ and $\gamma = 60°$) [3.39].

From experimental information it has become clear that additional configuration restrictions, other than those related to parity and signature, are needed to study various excited bands. For example, in superdeformed rotational bands the observed bands may well be specified in terms of the number of nucleons in high-N intruder shells [3.52]. Since N is not a good quantum number, calculational procedures have been developed to construct diabatic single-particle levels [3.31,39,51]. In practice, this means that "small" interactions between single-particle levels are removed. This leads to calculated configurations that do not show any drastic changes in their wave function, and configurations may be specified, for example, by the number of neutrons and protons in each N-shell.

Except for $\omega = 0$, no pairing calculations are performed in the NICRA code. Pairing is very important for the description of low-spin states, especially for those with 0, 1, or 2 quasiparticles excited. Customarily, low- to medium-spin spectra are analyzed with the aid of quasi-particle diagrams, as described in Ref. [3.53], but numerous more advanced calculations are also described in the literature. It is our experience, however, that a decent description of bands with more than two exited quasiparticles is usually obtained without invoking pairing.

A part of the total nuclear energy that has not received much attention in high-spin calculations is the macroscopic energy. Here also a number of refinements have been suggested. For example, the deformed-liquid-drop energy may be replaced by that of the droplet model, and also the finite range of the nuclear force may be taken into account (see Ref. [3.48] and references therein). Furthermore, one may introduce a diffuseness term in the expression for the rigid-body inertia.

The possibility of a deformation-dependent spin–orbit force was discussed in Ref. [3.54], and with the recent experimental achievements in nuclear structure at large deformations, interesting possibilities open up to test or fit the consequences of ε-dependent κ-parameters. To perform such a study would be quite straightforward within the present NICRA code.

It is in principle possible to increase the agreement with measured spectra by simply increasing the number of parameters to fit. One possibility is to introduce separate κ- and μ-parameters for each ℓ-shell. Another possibility is to introduce higher-order terms in the modified-oscillator potential like $\bar{\ell}^4$- and $\bar{\ell}^2(\bar{\ell} \cdot \bar{s})$-terms. Some efforts along these lines have been taken in Refs. [3.55,56]. Several ways to solve the problem caused by the velocity dependence in the $\bar{\ell}^2$-term have been discussed in the literature (see e.g. [3.15,57]). The suggested changes are easily incorporated in the NICRA code.

Acknowledgement

The Lund Nilsson–Strutinsky code has been developed over several years. For important contribution we especially want to thank S.G. Nilsson, B. Nilsson, P. Möller, S.E. Larsson, G. Leander, and R. Bengtsson. This project is supported by the Swedish Natural Science Research Council.

References

[3.1]   A. Bohr, Mat. Fys. Medd. Dan. Vid. Selsk. **26** (1952) No. 14

[3.2]   A. Bohr and B.R. Mottelson, Mat. Fys. Medd. Dan. Vid. Selsk. **27** (1953) No. 16

[3.3]   S.G. Nilsson, Mat. Fys. Medd., Dan. Vid. Selsk. **29** (1955) No. 16

[3.4]   V.M. Strutinsky, Nucl. Phys. **A95** (1967) 420

[3.5]   V.M. Strutinsky, Nucl. Phys. **A122** (1968) 1

[3.6]   S.G. Nilsson, C.F. Tsang, A. Sobiczewski, Z. Szymanski, S. Wycech, C. Gustafsson, I.-L. Lamm, P. Möller, and B. Nilsson, Nucl. Phys. **A131** (1969) 1

[3.7]   M. Bolsterli, E.O. Fiset, J.R. Nix, and J.L. Norton, Phys. Rev. **C5** (1972) 1050

[3.8]   M. Brack, J. Damgaard, A.S. Jensen, H.C. Pauli, V.M. Strutinsky, and C.Y. Wong, Rev. Mod. Phys. **44** (1972) 320

[3.9]   P. Möller, Nucl. Phys. **A192** (1972) 529

[3.10] S.E. Larsson, Phys. Scripta **8** (1973) 17

[3.11] G. Leander and S.E. Larsson, Nucl. Phys. **A239** (1975) 93

[3.12] R. Bengtsson, S.E. Larsson, G. Leander, P. Möller, S.G. Nilsson, S. Åberg, and Z. Szymanski, Phys. Lett. **57B** (1975) 301

[3.13] K. Neergård and V.V. Pashkevich, Phys. Lett. **59B** (1975) 218

[3.14] K. Neergård, V.V. Pashkevich and S. Frauendorf, Nucl. Phys. **A262** (1976) 61

[3.15] G. Andersson, S.E. Larsson, G. Leander, P. Möller, S.G. Nilsson, I. Ragnarsson, S. Åberg, R. Bengtsson, J. Dudek, B. Nerlo-Pomorska, K. Pomorski, and Z. Szymanski, Nucl. Phys. **A268** (1976) 205

[3.16] I. Ragnarsson, T. Bengtsson, G. Leander, and S. Åberg, Nucl. Phys. **A347** (1980) 287

[3.17] T. Bengtsson and I. Ragnarsson, Phys. Scripta **T5** (1983) 165

[3.18] W. Nazarewicz, J. Dudek, R. Bengtsson, T. Bengtsson, and I. Ragnarsson, Nucl. Phys. **A435** (1985) 397

[3.19] P. Möller, S.G. Nilsson, and J.R. Nix, Nucl. Phys. **A229** (1974) 292

[3.20] R. Bengtsson, J. Dudek, W. Nazarewicz, and P. Olanders, Phys. Scripta **39** (1989) 196

[3.21] S. Åberg, H. Flocard, and W. Nazarewicz, Ann. Rev. Nucl. Part. Sci. **40** (1990)

[3.22] I. Ragnarsson, S.G. Nilsson, and R.K. Sheline, Phys. Rep. **45** (1978) 1

[3.23] I. Ragnarsson and R.K. Sheline, Phys. Scripta **29** (1984) 385

[3.24] R. Bengtsson, P. Möller, J.R. Nix, and J.-Y. Zhang, Phys. Scripta **29** (1984) 402

[3.25] S.G. Nilsson and I. Ragnarsson, Shapes and Shells in Nuclear Structure (Cambridge Univ. Press), to be published

[3.26] P. Quentin and H. Flocard, Ann. Rev. Nucl. Part. Sci. **28** (1978) 523

[3.27] P. Bonche, H. Flocard, and P.H. Heenen, Nucl. Phys. **A467** (1987) 115

[3.28] M. Brack and P. Quentin, Nucl. Phys. **A361** (1981) 35

[3.29] D.R. Inglis, Phys. Rev. **96** (1954) 1059

[3.30] C. Gustafsson, I.-L. Lamm, B. Nilsson, and S.G. Nilsson, Arkiv för Fysik **36** (1967) 693

[3.31] T. Bengtsson and I. Ragnarsson, Nucl. Phys. **A436** (1985) 14

[3.32] S.E. Larsson, I. Ragnarsson, and S.G. Nilsson, Phys. Lett. **38B** (1972) 269

[3.33] S.E. Larsson, G. Leander, I. Ragnarsson, and N.G. Alenius, Nucl. Phys. **A261** (1976) 77

[3.34] S.G. Rohozinski and A. Sobiczewski, Acta Phys. Pol. **B12** (1981) 1001

[3.35] W. Nazarewicz and P. Rozmej, Nucl. Phys. **A369** (1981) 396

[3.36] W.D. Myers and W.J. Swiatecki, Ark. Phys. **36** (1967) 343

[3.37] S.E. Larsson and G. Leander, Proc. Conf. on Physics and Chemistry of Fission, Rochester, NY, 1973 (IAEA, Vienna, 1974) vol. I, p. 177

[3.38] G. Leander, Nucl. Phys. **A219** (1974) 245

[3.39] I. Ragnarsson, S. Åberg, and R.K. Sheline, Phys. Scripta **24** (1981) 215

[3.40] P. Rozmej; preprint GSI-85-41; P. Rozmej, K. Böning, and A. Sobiczewski, Proc. XXIV Int. Winter Meeting on Nuclear Physics Bormio, Italy, 1986 publ. in Ricerca Scientifica ed Educazione Permanente, Supp. Nr. **49**, 567 (University of Milano 1986)

[3.41] R.K. Sheline, I. Ragnarsson, S. Åberg, and A. Watt, J. Phys. G **14** (1988) 1201

[3.42] J.-Y. Zhang, N. Xu, D.B. Fossan, Y. Liang, R. Ma, and E.S. Paul, Phys. Rev. **C39** (1989) 714

[3.43] I. Ragnarsson, Proc. XXIII School on Physics, Zakopane, Poland, 1988, part 1, p. 53 (Institute of Nuclear Physics and Jagiellonian University, Krakow 1988).

[3.44] T. Bengtsson, Nucl. Phys. **A512** (1990) 124

[3.45] J.C. Bacelar, M. Diebel, C. Ellegaard, J.D. Garrett, G.B. Hagemann, B. Herskind, A. Holm, C.-X. Yang, J.-Y. Zhang, P.O. Tjøm, and J.C. Lisle, Nucl. Phys. **A442** (1985) 509

[3.46] A.S. Jensen, P.G. Hansen, and B. Jonson, Nucl. Phys. **A431** (1984) 393

[3.47] D.G. Madland and J.R. Nix, Nucl. Phys. **A476** (1988) 1

[3.48] P. Möller, J.R. Nix, and W.J. Swiatecki, Nucl. Phys. **A492** (1989) 349

[3.49] I. Ragnarsson, Phys. Rev. Lett. **62** (1989) 2084

[3.50] T. Bengtsson, I. Ragnarsson, and S. Åberg, to be published

[3.51] T. Bengtsson, Nucl. Phys. **A496** (1989) 56

[3.52] T. Bengtsson, I. Ragnarsson, and S. Åberg, Phys. Lett. **208B** (1988) 39

[3.53] R. Bengtsson and S. Frauendorf, Nucl. Phys. **A327** (1979) 139

[3.54] I. Hamamoto and W. Ogle, Nucl. Phys. **A240** (1975) 54

[3.55] I. Ragnarsson, unpublished

[3.56] S. Åberg, MSc Dissertation, Math. Phys. Report 1975, Lund, unpublished

[3.57] S.-I. Kinouchi, Thesis, University of Tsukuba, Jan. 1988

[3.58] I. Hamamoto and B. R. Mottelson, Phys. Lett. **132B** (1983) 7

[3.59] N. Onishi, I. Hamamoto, S. Åberg, and A. Ikeda, Nucl. Phys. **A452** (1986) 71

4. The Random-Phase Approximation for Collective Excitations

G. Bertsch

4.1 Introduction

The random-phase approximation (RPA) is a theory of small-amplitude vibrations in the quantum many-body system. The name was coined in the first application of the method by Bohm and Pines [4.1] to the plasma oscillations of an electron gas. The theory is equivalent to a limit of time-dependent Hartree–Fock theory in which the amplitude of the motion is small. Consequently, the applicability of the theory is restricted to systems where a Hartree–Fock or some effective mean-field theory provides a good description of the ground state. Thus the theory is best for closed-shell nuclei. To the extent that open-shell nuclei can be described with Hartree–Fock–Bogoliubov theory, a corresponding generalization can be applied, the quasi-particle RPA. The first applications in nuclear physics were to collective quadrupole states in open-shell nuclei by Baranger [4.2] and to the collective octupole state in spherical nuclei by Brown, Evans, and Thouless [4.3]. Early applications to nuclear physics were partly phenomenological, owing to deficiencies in the interaction and numerical limitations in the size of the configuration space. With improved interactions and present computer resources, RPA has proven itself to be a robust theory, capable of predicting and describing in detail many properties of collective excitations.

There are two quite different mathematical representations of the RPA, namely the response-function formalism and the A, B-matrix formulation. These are equivalent in principle, but in practice one or the other may be more suited to the problem at hand. The response-function formalism allows excitations to be calculated in very large spaces of configurations – the computational effort is only linear in the number of configurations. However, the interaction must have a simple form, with very restricted possibilities for nonlocality. In particular, the exchange interaction can only be calculated approximately, in a zero-range approximation. In contrast, the matrix formulation puts no limitations on the interaction, but the computational effort is cubic in the number of configurations included in the space. The response-function formalism is particularly convenient for dealing with excitations in the continuum, where the states acquire a width due to the escape of particles. It is much harder to deal with the continuum by the matrix method. The program presented in this chapter has as its primary focus the resonances in the continuum, and it uses the response-function method. The program follows closely the one described in Ref. [4.4]. Other nuclear studies using this method may be found in Refs. [4.5–7]. The method has also been applied to the study of collective electron excitations in spherical atomic clusters [4.8]. The matrix method has been applied extensively to nuclear excitations as in [4.9] and in [4.10]. The latter calculations in particular are self-consistent and use an interaction that is fully realistic as to range and exchange nonlocality.

As mentioned, the response-function method relies on a simple representation of the interaction. In fact, the interaction must be expressed by a sum of separable terms. The numerical difficulty in the evaluation of the response function is controlled by the number of terms in the interaction; the computation requires inversion of a matrix whose dimensionality is that number. In the program included in this chapter, the interaction is a δ-function, represented on a discrete mesh in coordinate space. The separability of the interaction is obvious from its numerical representation as $\delta(r_1 - r_2) \sim \sum_i \delta(r_1 - R_i)\delta(r_2 - R_i)\Delta R$, where R_i are mesh points and ΔR is the mesh spacing. As one of the compromises in the present program, no dependence of the interaction on currents or on spin is provided. However, the isospin degree of freedom is fully treated by evaluating neutron and proton responses, which are coupled together by the neutron–proton interaction. Other response programs including more general interactions are described in Ref. [4.5]. It is also possible to construct the interaction in momentum space and apply the response technique there [4.11]. However, the coordinate-space representation seems well-suited for interactions of the Skyrme type that are popular in Hartree–Fock calculations.

4.2   Derivation

The RPA may be derived from the small-amplitude limit of the time-dependent Hartree–Fock equations in an external field. The time-dependent Hartree-Fock equations have the structure

$$i\frac{\partial}{\partial t}\phi_i(t) = \left(-\frac{\nabla^2}{2m} + V[\rho] + \lambda V_{\text{ext}}\cos\omega t\right)\phi_i(t). \tag{4.1}$$

Here ϕ_i are the single-particle wave functions which depend on both coordinate and time. In (4.1) V is a potential due to the interacting particles, which in principle is a nonlocal functional of the single-particle density matrix. V_{ext} is an external potential that drives the system at angular frequency ω. In this term the real number λ is a smallness parameter to help us keep track of the first-order deviations from the equilibrium solution. This equilibrium solution is the set of wave functions $\{\phi_i^0\}$ which satisfy the static Hartree–Fock equations,

$$\left(-\frac{\nabla^2}{2m} + V[\rho_0]\right)\phi_i^0 \equiv H_0\phi_i^0 = \epsilon_i\phi_i \tag{4.2}$$

where ϵ_i are the single-particle energies. The perturbed wave function is expanded in the small-amplitude limit as

$$\phi_i = e^{-i\epsilon_i t}(\phi_i^0 + \lambda \delta \phi_i). \tag{4.3}$$

We also need to expand the density to first order in λ. The ordinary density is defined by

$$\rho(r,t) = \sum_i \phi_i^*(r,t)\phi_i(r,t).$$

The expansion is

$$\rho = \rho_0 + \lambda \delta \rho + O(\lambda^2)\,,$$

where

$$\rho_0 = \sum_i \phi_i^{0*}\phi_i^0$$

and

$$\delta \rho = \sum_i (\phi_i^{0*}\delta \phi_i + (\delta \phi_i)^*\phi_i^0). \tag{4.4}$$

We only consider here the simple case where the potential is a function of the ordinary densities (including spin and isospin densities). Then the perturbed potential may be expressed to order λ as

$$V[\rho] = V[\rho_0] + \lambda \frac{\partial V}{\partial \rho}[\rho_0]\,\delta \rho + O(\lambda^2). \tag{4.5}$$

The first term is just the equilibrium Hartree–Fock potential, and the second term, calculated by differentiating the function V with respect to ρ, behaves like a residual interaction. Inserting (4.4) and (4.5) in (4.1) and equating the terms linear in λ, we obtain the equation

$$i\frac{\partial}{\partial t}\delta \phi_i + \epsilon_i \delta \phi_i = H_0\,\delta \phi_i + \frac{\partial V}{\partial \rho}\,\delta \rho\,\phi_i^0 + V_{\text{ext}}\cos \omega t\,\phi_i^0. \tag{4.6}$$

Next $\delta \phi_i$ is separated into its real and imaginary parts, and (4.6) is written as separate equations for the two parts. A solution to the inhomogeneous equations is obtained by taking $\delta \phi_i$ to have the time dependence

$$\delta \phi_i(t) = \cos \omega t\,\text{Re}\,\delta \phi_i + i\sin \omega t\,\text{Im}\,\delta \phi_i.$$

We use the same symbol ϕ to denote both time-dependent and time-independent functions, but this should cause no confusion. The equations satisfied by the (time-independent) $\delta \phi_i$ are then

$$\omega\,\text{Im}\,\delta \phi_i = [H_0 - \epsilon_i]\,\text{Re}\,\delta \phi_i + \frac{\partial V}{\partial \rho}\delta \rho \phi_i^0 + V_{\text{ext}}\phi_i^0$$

$$\omega\,\text{Re}\,\delta \phi_i = [H_0 - \epsilon_i]\,\text{Im}\,\delta \phi_i. \tag{4.7}$$

We eliminate $\text{Re}\delta\phi_i$ from the two equations to obtain

$$\text{Re}\delta\phi_i = [\omega^2 - (H_0 - \epsilon_i)^2]^{-1}[H_0 - \epsilon_i]\left(\frac{\partial V}{\partial \rho}\rho\phi_i + V_{\text{ext}}\phi_i\right). \qquad (4.8)$$

The reciprocal operator is simplified using

$$(A^2 - B^2)^{-1}B = \frac{1}{2}[(A - B)^{-1} - (A + B)^{-1}].$$

The equation for $\text{Re}\delta\phi$ is then

$$\text{Re}\delta\phi_i = -\frac{1}{2}([H_0 - \omega - \epsilon_i]^{-1} + [H_0 - \epsilon + \omega]^{-1})\left(\frac{\partial V}{\partial \rho}\delta\rho + V_{\text{ext}}\right)\phi_i^0. \quad (4.9)$$

We can eliminate $\delta\phi_i$ as a variable by multiplying (4.9) by ϕ_i^{0*} and summing over i, which is equal to half of $\delta\rho$. The resulting equation is expressed compactly as

$$\delta\rho = -G^0\left(\frac{\partial V}{\partial \rho}\delta\rho + V_{\text{ext}}\right),$$

where G^0 is defined in coordinate space as

$$G^0(r, r') = \sum_i \phi_i^{0*}(r)([H_0 - \epsilon_i - \omega]_{r,r'}^{-1} + [H_0 - \epsilon_i + \omega]_{r,r'}^{-1})\phi_i^0(r'). \quad (4.10)$$

This equation is formally solved for $\delta\rho$ as

$$\delta\rho = G^{\text{RPA}}V_{\text{ext}}, \qquad (4.11)$$

where

$$G^{\text{RPA}} = \left[1 + G^0\frac{\partial V}{\partial \rho}\right]^{-1}G^0. \qquad (4.12)$$

This formal expression is the basis of the response-function method for obtaining the excitations of the system. The operators are approximated in some suitable basis and the resulting matrix equations are solved by matrix inversion. As mentioned in the beginning, a coordinate-space representation is very useful, and is the basis in the numerical program.

We next make an angular-momentum decomposition of the functions appearing in the response. This permits the RPA to be calculated from independent radial equations for each multipole, provided the Hartree–Fock state is spherically symmetric. We define the multipolar interaction v_L and response G_L as follows:

$$\frac{\partial V(r)}{\partial \rho(r')} = \sum_{LM} v_L(r_1, r_2)Y_{LM}^*(\hat{r}_1)Y_{LM}(\hat{r}_2)/r_1 r_2,$$

$$G^0(r_1, r_2) = \sum_{LM} G_L(r_1, r_2) Y_{LM}^*(\hat{r}_1) Y_{LM}(\hat{r}_2)/(r_1 r_2)^2 ,$$

where the Y_{LM} are the usual spherical harmonics. The single-particle Green's function is similarly expanded in a basis of the $(ls)j$-coupled single-particle states. Then the formula for the independent particle response is

$$G_L(r, r', \omega) = \sum_{j,j'} \frac{(2j+1)(2j'+1)}{4\pi(2L+1)} \left(j'\frac{1}{2}j - \frac{1}{2}|L0\right)^2 \phi_j(r)^* \phi_j(r')$$

$$([H_0 - \epsilon_j - \omega]_{j';r,r'}^{-1} + [H_0 - \epsilon_j + \omega]_{j';r,r'}^{-1}) , \qquad (4.13)$$

where the sum over single-particle states j, j' is restricted to those with orbital angular momentum l and l' such that $L + l + l'$ is even. In this equation, the ϕ_j are radial wave functions multiplied by r.

The radial Green's function is most conveniently evaluated in the coordinate representation by the usual Green's function formula for second-order linear differential equations,

$$[H_0 - \epsilon]_{r,r'}^{-1} = \frac{u(r_<)w(r_>)}{W}. \qquad (4.14)$$

Here u, w are solutions of the radial differential equation that have appropriate boundary conditions, i.e. u must be regular at the origin and w must be a pure outgoing wave at infinity:

$$u(r) \to r^{L+1} \quad \text{for} \quad r \to 0,$$

$$w(r) \to \exp(ikr) \quad \text{for} \quad r \to \infty.$$

W in (4.14) is the Wronskian of the differential operator,

$$W = \frac{1}{2m} \left(w\frac{du}{dr} - u\frac{dw}{dr}\right).$$

The program calculates G_L^{RPA} as a matrix in coordinate space. The radial transition density associated with the external field is then calculated by the relation

$$\delta\rho_L(r) = \int dr' G_L^{RPA}(r, r', \omega) V_{\text{ext}}(r'). \qquad (4.15)$$

This is related to the usual transition density by

$$\delta\rho(\mathbf{r}) = \delta\rho_L(r) Y_{LM}(\hat{r})/r^2.$$

Finally, the external field is integrated over the transition density to obtain the response function to the external perturbation of the system,

$$G_L(V_{\text{ext}}, \omega) = \int dr \delta\rho_L(r) V_{\text{ext}}(r) = \int dr dr' V_{\text{ext}}(r) G_L^{RPA}(r, r', \omega) V_{\text{ext}}(r').$$

The real and imaginary parts of this function are the primary output quantities that are printed.

4.3   Strength Function and Transition Strengths

Information about the transition strengths between an initial state i and final states f is contained in the strength function

$$S(\omega) = \sum_f \langle i|V_{ext}|f\rangle^2 \delta(E_f - E_i - \omega).$$

The relation to the response-function is simply

$$S(\omega) = \frac{1}{\pi}\text{Im}G_L(V_{ext}, \omega). \tag{4.16}$$

This formula is often used in applications. For example, inelastic scattering in the continuum is conveniently calculated using (4.16) with V_{ext} being the transition field of the projectile, or some approximation to the projectile interaction [4.12]. The experimental strength function is often smoother than the RPA prediction, either because of experimental resolution or because of physical damping processes not described by the RPA. To facilitate comparison with smoothed strengths, it is useful to artificially add an imaginary part to the frequency. This is equivalent to convoluting the strength function with a Lorentzian. Otherwise, only the boundary condition on the single-particle Green's function, that particles be outgoing, provides an imaginary part to the response and a width to the resonances.

In the neighborhood of a resonance of the system, the response function may be expressed as

$$G(V_{ext}, \omega) \sim \langle i|V_{ext}|R\rangle^2 (\frac{1}{E_R + i\Gamma/2 - \omega}), \tag{4.17}$$

where E_R and $\Gamma/2$ are the real and imaginary parts of the frequency of the mode. Equation (4.17) is useful for extracting the properties of individual vibrational modes from the response. To find the squared matrix element for an individual resonance, one first calculates the response over a small frequency interval containing the mode in question. One may then either fit the imaginary part to the function $\text{Im}(E_R + i\Gamma/2 - \omega)^{-1} = \Gamma/2/((E_R - \omega)^2 + (\Gamma/2)^2)$ or simply integrate over an interval containing the resonance,

$$\langle i|V_{ext}|R\rangle^2 \sim \int_{E_R - \Delta}^{E_R + \Delta} S(\omega)d\omega.$$

The transition density associated with a resonance is defined in terms of the matrix element of the density operator as

$$\delta\rho = \langle i|\hat{\rho}|R\rangle.$$

It may be calculated from the above quantities by the equation

$$\delta\rho = \frac{G^{\text{RPA}} V_{ext}}{\langle i|V_{ext}|R\rangle}.$$

For inelastic scattering of hadronic projectiles, the transition strengths are often expressed in terms of the collective-model parameter β_L. This model assumes that the nucleons in the nuclear interior move incompressibly, so that the transition density is surface peaked. The transition density is taken to have the form

$$\delta\rho_L \sim \frac{\beta_L}{\sqrt{2L+1}} R_0 \frac{\mathrm{d}\rho_0}{\mathrm{d}r} Y_{LM} , \tag{4.18}$$

where R_0 is an appropriately defined nuclear radius. The magnitude of the transition density is set by β_L. Since the RPA transition densities of collective states follow (4.18) quite well, the β_L parameters of the transitions can be extracted from the multipole matrix element by relating it to the equivalent matrix element from eq. (4.18). This yields

$$\beta_L = \frac{4\pi\sqrt{2L+1}\langle i|r^L Y_{LM}|R\rangle}{(L+2)AR_0\langle r^{L-1}\rangle}. \tag{4.19}$$

4.4 Sum Rules and the Spurious State

The integrated transition strengths obey sum rules which are respected by the RPA theory. It is an important check on calculations to see how well the sum rules are satisfied numerically. Also, it is convenient to express transition strengths as a fraction of the appropriate sum rule, which eliminates any possible ambiguity in the definition of the strength functions or the normalization of the operator matrix elements.

For the Hamiltonians considered here in which the potential only depends on density, the sum rules may be expressed generally as

$$\int S(\omega)\omega \mathrm{d}\omega = \int \mathrm{d}^3 r \frac{(\nabla V_{\mathrm{ext}})^2}{2m} \rho_0.$$

In the special case of multipole fields, $V_{\mathrm{ext}} = r^L Y_{LM}(\theta, \phi)$, the sum rule becomes [4.13]

$$\int S_L(\omega)\omega \mathrm{d}\omega = \frac{L(2L+1)}{8\pi m} \int \mathrm{d}^3 r \, r^{2L} \rho_0 = \frac{L(2L+1)}{8\pi m} A\langle r^{2L-2}\rangle. \tag{4.20}$$

This equation applies both to isoscalar and isovector fields, i.e. fields in which V_{ext} acts on both neutrons and protons with a positive or negative phase relationship, respectively. For a pure proton field, the number of particles A in (4.20) should be replaced by Z. In the program, the strength function is calculated for some range of energy, and the total strength in that interval is compared with the multipole sum rule.

Finally, we mention the spurious-state phenomenon. The fact that the Hartree–Fock state is translationally degenerate means that a fictitious excitation can be created simply by displacing the Hartree–Fock wave functions by a small amount. In the RPA, this appears as an excitation at $\omega = 0$, having $L=1$. Thus a good test of complete consistency in the calculation is the presence of this zero-frequency mode. In practice, the consistency will not be perfect, but there will always be an isoscalar mode at a frequency close to zero.

The existence of the spurious state causes some difficulty with the dipole sum rule. Since the spurious state has practically all of the isoscalar dipole strength, the isoscalar and proton sum rules will be grossly violated unless the contribution from that state is included.

4.5 Numerics

At the core of the program is the computation of the radial single-particle Green's function (4.14) in the subroutine GREEN. Here the simplest possible second-difference algorithm is used to integrate the radial Schrödinger equation. This part of the program consumes relatively little time, so there is not much point in using fancier methods. A mesh in coordinate space of 0.1 to 0.25 fm is quite adequate to obtain accuracy of less than 0.1 MeV on the single-particle eigenenergies. The boundary condition on the Green's function is that it corresponds to an outgoing wave if the particle is unbound, and it vanishes at the coordinate space boundary if the particle is bound.

The input Hartree–Fock wave functions should have been calculated with the same numerical Hamiltonian as the Green's function for complete consistency; in fact the same algorithm should be used. In the program, a subroutine STATIC is provided which computes Woods–Saxon single-particle wave functions with the same algorithm. This subroutine requires as input the mass and charge of the nucleus and the quantum numbers of the occupied orbits. If the self-consistency is not strictly maintained, transition strength may appear at frequencies corresponding to transitions between occupied states. These Pauli-forbidden transitions in principle cancel out in the two terms in (4.10). The boundary condition for the particle Green's function in the subroutine GREEN is consistent with the particle spectrum of the subroutine STATIC.

The most time-consuming part of the calculation for light nuclei is the inversion of the polarization matrix, the first factor in (4.12). Since the matrix is complex, readily available library inversion routines cannot be used; a simple matrix-inversion subroutine, MATR, is included with the program. However, only the vector array DRHRPA, defined in (4.15), is needed in the program. If time is a real consideration, it is faster to obtain DRHRPA by directly solving the linear equations relating it to the vector array DRHOF. The mesh size for the matrix is set independently of the single-particle wavefunction mesh. The multipole strength function can be computed to about 10% accuracy with a mesh as coarse as 1 fm. Unless the external field is strongly varying, there is no useful gain in accuracy with mesh spacings

smaller than 0.5 fm. To give an example of the dimensionality, the response
for ^{208}Pb requires a radial interval extending to about 9 fm. With a 1
fm mesh spacing and remembering that neutrons and protons are treated
separately, the matrix size is 18 by 18.

The residual interaction should also be input consistently with the
Hartree–Fock Hamiltonian. However, to make the program run as is, a subroutine VRESID is included that constructs a residual interaction based on
the Skyrme parameterization. To alleviate the lack of consistency between
this subroutine and STATIC, a parameter VSCAL is provided to renormalize
the strength of the interaction. This may chosen to put the spurious state at
zero frequency. As the program stands, the required renormalization factor
is in the range 0.8–1.0.

The external field is determined in the subroutine EXTV. The spatial form
of the field is a pure multipole, $V_{ext} = r^L Y_{LM}$. There are three choices for
the isospin character of the field: it can act on protons only, on protons
and neutrons with equal amplitude (isoscalar), or on protons and neutrons
with opposite amplitudes (isovector). The sum rule for the field (4.20) is
calculated in the subroutine SUMRLE. Other fields besides multipole fields
may be of interest. For example, electron scattering could be represented
by a plane-wave field, which would have a Bessel-function radial dependence in the multipole expansion. If the user substitutes another field, the
corresponding sum rule is automatically calculated by SUMRLE.

A detailed description of the input format is provided in Table 4.1.

4.6   Tests and Studies

A data set is provided for the response of ^{16}O. The first thing to examine
is the consistency between the single-particle wave functions, the single-
particle Hamiltonian H_0, and the residual interaction. If these are self-
consistent, there will be a zero-frequency mode in the $L = 1$ response. The
sample data set is set up to make this computation. What follows are
some things to note from the output of this first test run. The free particle
response will show characteristic behavior due to the existence of particle-
hole excitations at a frequency corresponding to the shell-gap energy. In
^{16}O, the shell gap is about 14 MeV. The real part of the response is negative
just above the transition frequency, and is large and positive just below it,
as in (4.17). The imaginary part of the response has a peak just at the
transition frequency. Of course, if the states are bound, the imaginary
part will vanish, unless an imaginary component is imposed numerically on
the single-particle Green's function. This may be accomplished by giving
nonzero values to the parameter GAM.

The interacting response will have a peak near zero frequency, the spurious state. The real part of the response will be positive if the spurious
state has positive energy and negative if the residual interaction is so strong

Table 4.1. Input to the RPA program. The numerical values are the data appropriate for the calculation of the dipole response of the nucleus ^{16}O.

Line	Variable	Explanation
1	DEL, NGRID	DEL=0.25 fm is the mesh size for the Schrödinger equation and single-particle Green's function. NGRID=50 is the number of mesh points. Thus, wave functions are calculated over the spatial range 0 to DEL*NGRID=12.5 fm.
2	A,Z	A=16 and Z=8 are the mass and charge of the nucleus, necessary to determine an appropriate Woods–Saxon potential to construct the (not Hartree–Fock) single-particle wave functions.
3.1 ... 3.n	L,J,NQ, NODE	These are the quantum numbers for the occupied orbitals, with J twice the $(ls)j$-coupled angular momentum and NQ the charge. For example, the lowest $p_{3/2}$ proton orbital has L=1,J=3,NQ=1,NODE=0. The orbital data is terminated by the line -1 0 0.
4	N, DEL2	The mesh for the response function has N=10 points spaced by DEL2=0.5 fm.
5	T0,T3, X, VSCAL	These are parameters of the residual interaction in the Skyrme formulation, with VSCAL an overall scaling factor. Self-consistent values for infinite nuclear matter are close to T0=-1100 MeV fm^3, T3=15,000 MeV fm^5. The spin exchange parameter X has the value X=0.5. Because the self-consistency is not implicit in the wave functions, VSCAL=0.93 to produce a spurious state close to zero frequency.
6	L,EX, EXM, DEX, GAM	Here L=1 is the multipolarity of the response. The response is calculated on a grid of energies starting from EX=0 MeV and ranging up to EXM=40 MeV in steps of DEX=1 MeV. GAM=1 MeV is an added imaginary part of the energy (actually i*GAM/2 is added) to smooth the strength function.
7	I	I is a character specifying the isospin of the external field. I=0 and I=1 correspond to isoscalar and isovector external fields. For any other value, e.g. I=-1, the external field acts only on the protons.

as to bring it below zero frequency. Assuming that the input is not completely self-consistent, vary the interaction scaling parameter VSCAL to put the spurious state near zero. Also, it is interesting to see how sensitive the frequency of the lowest state is to the interaction.

As mentioned before, it is important to check the sum rule. For this one needs a large range of energies, and one needs to sample the energies at small enough intervals to pick up all the states. Examine the proton L=1 response in ^{16}O, using several values of GAM and a step size of 1 MeV in the mesh of frequencies. Also, set VSCAL slightly smaller than the self-consistent value so that the spurious state in included in the sum. Why does GAM=0 give poor results? How large a GAM is required to smooth the strength function sufficiently for the 1 MeV frequency mesh? How high an energy interval must be included to get 90% of the sum rule?

Examine the isoscalar quadrupole response by setting L=2 in data line 6 and I=0 in line 7. You will see that the attractive interaction causes the quadrupole strength to be shifted down from the independent particle strength function. In ^{16}O, the RPA predicts that all of the strength is in a peak at about 20 MeV excitation, the giant quadrupole resonance. The energy agrees well with the empirical value of 22 MeV [4.14]. What is the width of the resonance? The RPA width will be small compared to the empirical width, which is of the order of 6 MeV. This is due to the neglect in RPA of damping into more complicated configurations than those having a 1-particle 1-hole structure. Do the same for the nucleus ^{208}Pb. Here the giant quadrupole is at 10.9 MeV, and furthermore there is significant quadrupole strength at lower excitation. What fraction of the sum rule is in the lower portion of the strength function? Experimentally, there is a 2^+ state at 4.086 MeV which has a strength given by $\beta_2 = 0.058$ [4.15]. Using (4.19), compare the predicted strength with the empirical.

It is also interesting to see how close the transition density is to the collective model. To get the radial transition density from the program, it is necessary to print out the vector array DRHRPA. It may then be compared with the collective model shape, $\frac{d\rho_0}{dr}$. Remember that the radial transition density includes an extra factor of r^2.

If a nucleus is unstable with respect to quadrupole deformation in the Hartree–Fock ground state, the L=2 RPA response of the spherical state will be negative at zero frequency (owing to the presence of an imaginary eigenfrequency). Are there any j-shell closures that are unstable with respect to quadrupole deformation? For example, the nucleus ^{28}Si with a filled $d_{5/2}$ shell is on the borderline of stability. If the strength of the spin–orbit potential in the single-particle Hamiltonian is decreased, it becomes deformed.

The isoscalar monopole resonance (L=0) is very interesting because of its close connection with the compression modulus of nuclear matter [4.10]. What is the predicted frequency in ^{208}Pb of this mode? Experimentally, it lies at 14 MeV [4.16]. The RPA prediction with Skyrme interactions is at a higher excitation energy. The Skyrme interaction apparently has too strong a density dependence. Change the interaction (via VSCAL) to put the monopole resonance at the correct excitation. What is its width? This should be compared with the experimental width of $\sim$ 3 MeV.

There is a 0^+ state in ^{4}He at 20.1 MeV excitation having a width of 0.27 MeV [4.17]. Can this state be reasonably described as a monopole resonance? Although the use of RPA for a 4-particle nucleus is problematic, it is interesting to see whether the width is consistent with the monopole assumption. To make a fair test, the input parameters in the program need to be adjusted to give the correct separation energy of the ^{4}He nucleus as well as reproduce the energy of the monopole.

Examine the octupole response of the nucleus ^{208}Pb. The lowest L=3 state in this nucleus has a transition strength of 38 single-particle units, making it one of the most collective octupole states of any nucleus. The state has an excitation energy of 2.6 MeV. The square of its charge matrix element $er^2 Y_{LM}$ is given by the $B(E3, 3 \rightarrow 0)$, with an experimental value $B(E3, 3 \rightarrow 0) = 0.9 \times 10^5$ e^2 fm^6 [4.15]. Its β moment is 0.12. Determine the theoretical position and strength of this state, being sure that VSCAL is set to satisfy the consistency condition on the dipole state. It is also interesting to compare the shape of the transition density of this state with experiment [4.18].

The giant dipole resonance may be examined using the isovector field, setting the parameter I=1 in line 7. Determine where the center of gravity of the strength is for ^{16}O. Experimentally, it is much higher, at about 25 MeV excitation. The disagreement is partly due to an inadequate description of the interaction in the program. We have assumed that the potential field depends only on local density. In fact, the Hartree–Fock field is nonlocal, so that nucleons in a nucleus have an effective mass of about 0.75 times the free mass. This is not noticeable in the isoscalar motion, because when the nucleons move together the interaction restores the potential field to give back the free mass. An ad hoc way to circumvent this deficiency is to change the mass of the nucleon by hand in the program. Put in an effective mass of 0.75 the free mass, and see how well the giant dipole is reproduced. What is the width of the predicted dipole resonance? The experimental width of the dipole in light nuclei is attributed to the nucleon escape width, and so should be reproduced in the model. Try the same study in ^{208}Pb. Here the giant dipole resonance is at 13.5 MeV; it has a width of 4 MeV and has a very smooth shape [4.19]. In this case, the escape width is only a small fraction of the total width and the shape of the RPA resonance bears little resemblance to the actual line shape of the giant dipole resonance.

Are there threshold effects in the dipole response associated with nearly unbound s-wave orbitals? There have been speculations that nuclei near the limit of zero nucleon binding may have unusually strong dipole transition strengths at low excitation. See for example Ref. [4.20]. Calculate the dipole response for a nucleus that is very neutron rich, such as ^{10}He (which does not exist in nature) or ^{28}O. Is there significant strength just above the particle emission threshold?

4.7　Technical Note

The program expects input to be read from the keyboard. Example values for the input parameters as well as a detailed description of the input can be found in Table 4.1.

Acknowledgement

The author thanks W. Bauer and A. Bulgac for careful reading of the manuscript. This work was supported by the National Science Foundation under Grant 87-14432.

References

[4.1] D. Pines and D. Bohm, Phys. Rev. **85** (1952) 338

[4.2] M. Baranger, Phys. Rev. **120** (1960) 957

[4.3] G.E. Brown, J.A. Evans, and D.J. Thouless, Nucl. Phys. **24** (1961) 1

[4.4] S. Shlomo and G. Bertsch, Nucl. Phys. **A243** (1975) 507

[4.5] G. Bertsch and S.F. Tsai, Physics Reports **18** (1975) 126

[4.6] K.F. Liu and G.E. Brown, Nucl. Phys. **A265** (1976) 385

[4.7] N. Van Giai and H. Sagawa, Nucl. Phys. **A371** (1981) 1

[4.8] W. Ekardt, Phys. Rev. **B31** (1985) 6360

[4.9] P. Ring and J. Speth, Nucl. Phys. **A235** (1974) 315

[4.10] J.P. Blaizot, D. Gogny, and B. Grammaticos, Nucl. Phys. **A265** (1976) 315; Physics Reports **64** (1980) 171

[4.11] W. Knuepfer and M.G. Huber, Phys Rev. **C14** (1976) 2254; Z. Phys. **A276** (1976) 99

[4.12] S.F. Tsai and G. Bertsch, Phys. Rev. **C11** (1975) 1634

[4.13] A.M. Lane, *Nuclear Theory* (Benjamin, 1964) 125

[4.14] F. Bertrand, Ann. Rev. Nucl. Sci. **26** (1976) 457

[4.15] Nuclear Data Sheets **47** (1986) 840

[4.16] J. Speth and A van der Woude, Reports on Progress in Physics **44** (1981) 719

[4.17] S. Fiarman and W.E. Meyerhof, Nucl. Phys. **A206** (1973) 1

[4.18] J. Heisenberg, Advances in Nuclear Physics **12** (1981) 61

[4.19] B. Berman and S. Fultz, Reviews of Modern Physics **47** (1975) 713

[4.20] T. Uchiyama and H. Morinaga, Z. Phys. **A320** (1985) 273

5. The Program Package PHINT for IBA Calculations

Olaf Scholten

5.1 Purpose

The program package "PHINT" contains programs to perform calculations of even–even nuclei in the framework of the IBA model. The programs are coded in FORTRAN. The available programs are:

Program name	Calculates
PCIBAXW	Excitation energies and wave functions
PCIBAEM	Electromagnetic transitions
CFPGEN	Coefficients of fractional parentage generator

5.2 Introduction

The present program package PHINT has been written to perform calculations in the interacting-boson-approximation (IBA) model [5.1–9]. In this model low-lying collective states of medium-heavy nuclei ($A > 100$) are described by a system of interacting bosons. In the IBA model, two different bosons are considered: the s-boson, which carries zero angular momentum, and the d-boson, which is an angular-momentum-2 state. What distinguishes the IBA model from most other boson models is the fact that the total number of bosons (s plus d) is strictly conserved and finite [5.5]. This last property makes it possible to calculate the full Hamiltonian matrix and diagonalize it without the need for model space truncations. Therefore, the properties of a finite system of mutually interacting bosons can be studied exactly – within the limits of numerical accuracy – without any restriction on the two-body interaction.

In the most commonly used microscopic picture of the IBA model [5.6,7] the bosons are considered to be the equivalent of pairs of nucleons. The importance of the s-boson is the result of the strong monopole pairing component in the nucleon–nucleon force between like particles. The importance of the d-boson lies in the moderately strong quadrupole pairing force and in the fact that the effective neutron–proton interaction is dominated by the quadrupole–quadrupole force. The microscopic background of the model is most easily taken into account in the so-called IBA-2 model where neutron and proton bosons are considered explicitly. In the present programs calculations are done in the simpler IBA-1 model, where no distinction is made between the two kinds of nucleon. The model space can be considered the subspace of that of IBA-2 consisting of those states which are fully symmetric under the interchange of neutron and proton indices. For low-lying states, the IBA-1 model has proven to be quite adequate [5.8].

For certain nontrivial values of the model parameters, the IBA Hamiltonian, the calculated spectrum, and the transition rates show some simple, regular structures. In these cases the Hamiltonian has a dynamical symmetry and the spectrum can also be calculated using analytical formulas [5.1–3] (see Sect. 5.5). Collective nuclei can be divided into three categories: (i) vibrators [5.1], (ii) axially symmetric rotors [5.2] and (iii) triaxial or gamma-unstable rotors [5.3]. For each of these categories there exists a dynamical symmetry that has a spectrum of similar structure. One of the strong points of the IBA model is that, in a numerical calculation, it is not only these limiting cases or symmetries that can be reproduced, but also intermediate ones [5.9].

5.3   Possible Applications

The present program package can be used for several different applications. Most of these are based on the fact that the solutions of a model Hamiltonian can be obtained for a continuous set of parameter values, both for very particular cases where nontrivial analytic solutions exist, and for those (possibly only minutely different) where such solutions are no longer available. A few possible applications are:

- The calculation of properties of collective nuclei. The IBA model was developed for this purpose. With these programs, the spectra and transition rates of a great variety of collective nuclei can be calculated. Most of the examples given in this chapter deal with this application and will thus not be elaborated on here.
- The investigation of "phase transitions" in finite boson systems. It can be shown that, in the limit of infinite boson number, first- and second-order phase transitions occur between the different dynamical symmetries [5.10]. In finite systems these abrupt changes in the properties are smoothed out (depending on the number of bosons), but even with only seven bosons they can be clearly recognized. Since the model Hamiltonian is solved exactly (within the limits of numerical accuracy) for all energy levels, level densities, spacings, and correlation lengths (transition energies) can be studied near the phase transition.
- The investigation of chaos in quantum systems [5.11]. Near a dynamical symmetry the energy spectrum shows all kinds of regularity and is certainly not chaotic. In general, for an arbitrary set of parameters, the calculated spectrum shows no regular structure whatsoever and could be classified as chaotic. The questions that could be addressed are: at what point does a quantum system behave chaotically and what are the most useful characteristic quantities?
- The test of the accuracy of certain approximation schemes. For example, the accuracy of perturbation schemes can be tested against the

results of the exact numerical calculation for any value of the perturbation parameter.

This list is certainly not intended to be exhaustive and many other topics can be added.

5.4   Description of the Programs

5.4.1   The Program "PCIBAXW"

The computer program "PCIBAXW" calculates energies and eigenvectors for positive and negative-parity states in the framework of the IBA model [5.4]. The negative parity states are constructed through the inclusion of one $L^\pi = 3^-(f)$ boson. The energies and eigenvectors are calculated by an exact diagonalization of the full IBA Hamiltonian, giving all the eigenvalues and an optional number of eigenvectors.

In calculating the matrix elements of the Hamiltonian the spherical (SU(5)) basis [5.1] is used:

$$|\psi\rangle = |[N], n_d, n_\beta, n_\Delta, L_d, n_f, L\rangle, \tag{5.1}$$

where

N = total number of bosons = $n_s + n_d$ ($+n_f$),
n_d = number of d-bosons,
n_β = number of pairs of d-bosons coupled to $L = 0$,
n_Δ = number of triplets of d-bosons coupled to $L = 0$,
L_d = total angular momentum of d-bosons,
n_f = number of f-bosons; =0 for positive parity states
                                               =1 for negative parity states,
L =total angular momentum of the state.

The natural form for the Hamiltonian in the SU(5) basis [5.1] is

$$H = H_{sd} + H_f + V_{df}, \tag{5.2}$$

with

$$
\begin{aligned}
H_{sd} =\ & \mathrm{HBAR}\, n_d \\
& + \sum_{L=0,2,4} \tfrac{1}{2}\sqrt{2L+1}\, C(\tfrac{L+2}{2})\, [(d^\dagger d^\dagger)^{(L)} \times (\tilde{d}\tilde{d})^{(L)}]_0^{(0)} \\
& + F\left([(d^\dagger d^\dagger)^{(2)} \times (\tilde{d}s)^{(2)}]_0^{(0)} + \text{h. c.}\right) \\
& + G\left([(d^\dagger d^\dagger)^{(0)} \times (ss)^{(0)}]_0^{(0)} + \text{h. c.}\right) \\
& + \tfrac{1}{2}\mathrm{CH1}\, [s^\dagger s^\dagger ss]_0^0 + \mathrm{CH2}\, \sqrt{5}[(d^\dagger s^\dagger)^{(2)} \times (\tilde{d}s)^{(2)}]_0^{(0)},
\end{aligned}
\tag{5.3}
$$

$$H_f = \text{EPS3 } n_f, \tag{5.4}$$

$$\begin{aligned} H_{df} = & \sum_{L=1}^{5} \sqrt{2L+1} \; D(L) \; [(d^\dagger f^\dagger)^{(L)} \times (\tilde{d}\tilde{f})^{(L)}]_0^0 \\ & -\text{F3} \; ([(d^\dagger f^\dagger)^{(3)} \times (f\tilde{s})^{(3)}]_0^0 + \text{h. c.}) + \text{EPSD } n_d n_f \; . \end{aligned} \tag{5.5}$$

The parameters HBAR, $C(I = \frac{L+2}{2})$, F, G, CH1, CH2, HBAR3, D(L), F3, and EPSD refer to the variable names as used in the program. To calculate the positive-parity states only, H_f and H_{df} are unimportant.

To specify the parameters for the Hamiltonian it is possible to use the *completely equivalent* "multipole expansion" of the Hamiltonian

$$\begin{aligned} H_{sd} = & \text{ EPS } n_d + \tfrac{1}{2}\text{ELL} \, (\boldsymbol{L} \cdot \boldsymbol{L}) + \tfrac{1}{2}\text{QQ}(\boldsymbol{Q} \cdot \boldsymbol{Q}) \\ & -5\sqrt{7} \text{ OCT } [(d^\dagger \tilde{d})^{(3)} \times (d^\dagger \tilde{d})^{(3)}]_0^{(0)} \\ & +15 \text{ HEX } [(d^\dagger \tilde{d})^{(4)} \times (d^\dagger \tilde{d})^{(4)}]_0^{(0)}, \end{aligned} \tag{5.6}$$

where

$$\boldsymbol{L} \cdot \boldsymbol{L} = -10\sqrt{3} \, [(d^\dagger \tilde{d})^{(1)} \times (d^\dagger \tilde{d})^{(1)}]_0^{(0)} \tag{5.7}$$

and

$$\begin{aligned} \boldsymbol{Q} \cdot \boldsymbol{Q} = & \sqrt{5}\Big[\{(s^\dagger \tilde{d} + d^\dagger s)^{(2)} + \frac{\text{CHQ}}{\sqrt{5}}(d^\dagger \tilde{d})^{(2)}\} \\ & \times \{(s^\dagger \tilde{d} + d^\dagger s)^{(2)} + \frac{\text{CHQ}}{\sqrt{5}}(d^\dagger \tilde{d})^{(2)}\}\Big]_0^{(0)}. \end{aligned} \tag{5.8}$$

It is possible to make a similar multipole expansion for H_{df}. In this case, however, only three terms in the expansion are used in the program,

$$H_{df} = \text{FELL} \, (\boldsymbol{L}_d \cdot \boldsymbol{L}_f) + \text{FQQ} \, (\boldsymbol{Q}_d \cdot \boldsymbol{Q}_f) - \text{KAP3} \, (O^{(3)} \cdot O^{(3)}), \tag{5.9}$$

where

$$\boldsymbol{L}_d \cdot \boldsymbol{L}_f = -2\sqrt{210} \, [(d^\dagger \tilde{d})^{(1)} \times (f^\dagger \tilde{f})^{(1)}]_0^{(0)}, \tag{5.10}$$

$$\boldsymbol{Q}_d \cdot \boldsymbol{Q}_f = -2\sqrt{35} \Big[\{(s^\dagger \tilde{d} + d^\dagger s)^{(2)} - \frac{\text{CHQ}}{\sqrt{5}}(d^\dagger \tilde{d})^{(2)}\} \times (f^\dagger \tilde{f})^{(2)}\Big]_0^{(0)} \tag{5.11}$$

and

$$O^{(3)} = (s^\dagger \tilde{f} + f^\dagger s)^{(3)} + \chi_3(d^\dagger \tilde{f} + f^\dagger \tilde{d})^{(3)}. \tag{5.12}$$

In the formulas (5.6–12), "EPS, PAIR, ELL, QQ, ECT, HEX, FELL, FQQ, CHQ, KAP3" refer to variable names as used in the program.

Another alternative is to use the IBA-2 formalism [5.7,13] to specify the parameters of the Hamiltonian. In IBA-2 the neutron and proton degrees of freedom are treated explicitly. This has the advantage of being closer

to a microscopic theory [5.6,7]. The matrices that have to be diagonalized
are, however, much larger. One can regard the IBA-1 model space, in which
neutron and proton degrees of freedom are not distinguished, as a subspace
of the IBA-2 model space, namely that of the fully symmetric states. From
the IBA-2 Hamiltonian one can thus project out its IBA-1 piece [5.8]. In the
present context the relevant terms in the IBA-2 Hamiltonian are

$$
\begin{aligned}
H_{NP} \; = \; & \mathrm{ED}(n_{d_\nu} + n_{d_\pi}) + \mathrm{RKAP}\, \boldsymbol{Q}_\nu^{(2)} \cdot \boldsymbol{Q}_\pi^{(2)} \\
& + \sum_{L=0,2,4} \tfrac{1}{2}\sqrt{2L+1}\,\mathrm{CLN}(I)[(d_\nu^\dagger d_\nu^\dagger)^{(L)}(\tilde{d}_\nu \tilde{d}_\nu)^{(L)}]^{(0)} \\
& + \sum_{L=0,2,4} \tfrac{1}{2}\sqrt{2L+1}\,\mathrm{CLP}(I)[(d_\pi^\dagger d_\pi^\dagger)^{(L)}(\tilde{d}_\pi \tilde{d}_\pi)^{(L)}]^{(0)},
\end{aligned}
\tag{5.13}
$$

where

$$
\boldsymbol{Q}_\rho^{(2)} = (s^\dagger \tilde{d} + d^\dagger s)_\rho^{(2)} + \chi_\rho (d^\dagger \tilde{d})_\rho^{(2)}, \qquad \rho = \nu, \pi.
\tag{5.14}
$$

In the input χ_ν is denoted by CHN and χ_π by CHP. (The parameters
appearing in (5.13) and (5.14) are the same as those appearing in the input
of the program NPBOS [5.13]. In NPBOS more parameters appear but they
are mostly related to the antisymmetric states.) In the projection from IBA-
2 onto IBA-1 the numbers of neutron (N_ν) and proton (N_π) bosons play an
important role. In the output they are given as NN and NP.

If in the IBA-2 calculation (with the program NPBOS) the lowest states
are indeed fully symmetric, the calculation with PCIBAXW, using the projection
procedure described above, will give exactly the same excitation energies.
Owing to admixtures of non-fully-symmetric states in IBA-2, the projection
gives different results and the parameters, mostly ED and RKAP, have to be
renormalized.

In its standard version the program can handle up to 7 d-bosons. This
limit, however, can easily be changed (see program "CFPGEN"). To calculate
the matrix elements of the Hamiltonian, the program uses the coefficients
of fractional parentage for the d-bosons (CFP) [5.12]. These CFP are read
in from a file "PHINT.CFP", which is produced by the program "CFPGEN".
The eigenvectors resulting from the diagonalization can be written in a file
"PHINT.WAV". These are necessary for the calculation of transition probabilities (see program "PCIBAEM"). The condensed output will appear on the
screen. A more complete version is written on the file "PCIBAXW.OUT".

Input for "PCIBAXW"

The appropriate form of the Hamiltonian should first be selected (e.g., by
comparing the nuclear spectrum to the analytically solvable limiting cases).
The parameters correspond to those of (5.3), (5.6), or (5.13). A full list with
default values is given in Table 5.1. After entering the parameters the user

is asked if the values are indeed correct. If not, there is another opportunity for entering them. The default values are now those given previously.

After the parameters for the d-boson Hamiltonian are set, there is the possibility for coupling the f-boson. The parameters for the d–f interaction are specified by (5.5) if (5.3) is used for the d-boson Hamiltonian, otherwise (5.9) is used.

Example of a "PCIBAXW" run

Input:
(Italic: response of the program, roman: input)
Explanatory information is bracketed, [R] = return key.

Hamiltonian parameters, use IBA-2 projection? (n/y) N [R]
Number of bosons = 7 [R]
Hamiltonian parameters, use multipole form? (y/n) Y [R]
Multipole form will be used.
EPS = 0.5 [R]
ELL = [R]
(When only a Return is given, the default value is taken)
QQ = −0.1 [R]
CHQ = [R]
OCT = [R]
HEX = [R]
EPS= 0.5000 , ELL = 0.0000 , QQ =−0.1000 , CHQ =−2.9580
OCT= 0.0000 , HEX= 0.0000
Are these parameters OK? y/n [R]
(The default answer on a Y/N question is Y)
Should negative parity states be calculated? n/y [R]
(The default answer on a N/Y question is N)
Print eigenvectors? n/y [R]
Print Hamiltonian matrix? n/y [R]
Print probability distribution for eigenvectors? n/y Y [R]
How many eigenvectors per spin, default = 4
NEIG = [R]

Output:

See output file "PCIBAXW.OUT". The total number of bosons (s and d) included in the calculation is 7 (=NPMMAX), including the full set of d-boson numbers up to 7. There is therefore no basis truncation. This truncation is only activated when a boson number is requested for which no CFPs have been calculated. The next lines give the parameters of the Hamiltonian, using (5.3–5,9). The printout discussed so far will always appear. When a printout of the eigenvectors is requested (not in this example), the basis

Table 5.1. Input parameters with their default values.

Variable	Default Value	Meaning
d-boson Hamiltonian according to (5.6)		
EPS	0.	d-boson energy
ELL	0.	strength of $L \cdot L$-force
QQ	0.	strength of $Q \cdot Q$-force
CHQ	$-\sqrt{35}/2$	parameter in quadrupole operator
OCT	0.	strength of octupole force
HEX	0.	strength of hexadecupole force
d-boson Hamiltonian according to (5.13)		
ED	0.	d-boson energy
RKAP	0.	strength of neutron–proton quadrupole force
CHN	0.	parameter in neutron quadrupole operator
CHP	0.	parameter in proton quadrupole operator
CLN	0.,0.,0.	d-boson conserving anharmonicity
CLP	0.,0.,0.	d-boson conserving anharmonicity
d-boson Hamiltonian according to (5.3)		
HBAR	0.	d-boson energy
C	0.,0.,0.	d-boson conserving anharmonicity
F	0.	one d-boson changing anharmonicity
G	0.	two d-boson changing anharmonicity
CH1	0.	see (5.3)
CH2	0.	see (5.3)
f-boson Hamiltonian (5.4)		
EPS3	0.	f-boson energy
d–f boson interaction according to (5.5)		
D	0.,0.,0.,0.	d-boson conserving anharmonicity
F3	0.	one d-boson changing anharmonicity
EPSD	0.	see (5.5)
d–f boson interaction according to (5.9)		
FELL	0.	strength of $L \cdot L$-force
FQQ	0.	strength of $Q \cdot Q$-force
KAP3	0.	strength of octupole force
CHO	0.	parameter χ in octupole operator

vectors are printed using the convention of (5.1). The total number of bosons, N, however, is not printed each time. The eigenvectors of the first four states are printed as column vectors.

Below the eigenvalues the probability distributions of the eigenvectors are printed as row vectors. The values printed are the squares of the amplitudes, summed over the basis states with the same number of d-bosons. This should be read as: the first eigenvector has a probability of 6.08% in the subspace of 0 d-bosons, 0% in the space of one boson, 30.66% in the space of 2 d-bosons, etc.

The "binding energy" printed at the end of the output is the eigenvalue of the first 0^+ state. This equals the binding energy (in MeV) up to a function, depending at most quadratically on the total number $N = N_s + N_d$ of bosons. The value of "EPS−EFF" is the difference between the effective energies of the s- and d-bosons. Normally, this quantity should be positive.

5.4.2 The Program "PCIBAEM"

This program calculates electromagnetic transition rates. The possible transitions and corresponding operators are [5.1]:

$$\hat{T}(E0) = \text{E0} \sqrt{5}(d^\dagger \tilde{d})_0^{(0)} = \text{E0} \, \hat{n}_d \, , \tag{5.15}$$

$$\hat{T}(M1) = [\text{M1} + \text{M1ND} \times \hat{n}_d] \, \hat{L} + \text{M1Q} \, [\hat{T}(E2) \times \hat{L}]^{(1)} \, , \tag{5.16}$$

$$\hat{T}(E2) = \text{E2SD}(s^\dagger \tilde{d} + d^\dagger s)^{(2)} + \frac{1}{\sqrt{5}} \text{E2DD} \, (d^\dagger \tilde{d})^{(2)} \, , \tag{5.17}$$

$$\hat{T}(E4) = \frac{1}{\sqrt{9}} \text{E4DD}(d^\dagger \tilde{d})^{(4)} \, , \tag{5.18}$$

$$\hat{T}(E1) = \text{E1Q} \, [\hat{T}(E2) \times (s_f^\dagger \tilde{f} + f^\dagger \tilde{s}_f)^{(3)}]^{(1)}$$
$$+ \text{E1DF} \, (d^\dagger \tilde{f} + f^\dagger d)^{(1)} \, , \tag{5.19}$$

$$\hat{T}(E3) = \text{E3Q} \, [\hat{T}(E2) \times \{(s_f^\dagger \tilde{f} + f^\dagger s_f)^{(3)}\}]^{(3)}$$
$$+ \text{E3} \, (s_f^\dagger \tilde{f} + f^\dagger s_f)^{(3)} + \text{E3DF} \, (d^\dagger \tilde{f} + f^\dagger d)^{(3)} \, . \tag{5.20}$$

Besides the input, discussed below, the program needs as input "PHINT.CFP", with the CFP, and "PHINT.WAV", with the necessary information on the eigenvectors, as produced by the program PCIBAXW. The output of the program is written onto the screen and the file "PCIBAEM.OUT".

The quadrupole moment for a state with angular momentum L is defined as

$$Q_L = \sqrt{\frac{16\pi}{5}} \sqrt{\frac{L(2L-1)}{(L+1)(2L+3)}} \sqrt{B(E2; L \to L)}, \tag{5.21}$$

where $\sqrt{B(E2; L \to L)}$ stands for

$$\sqrt{B(E2; L \to L)} = \langle L||T(E2)||L\rangle/\sqrt{2L+1}. \tag{5.22}$$

This quantity carries the sign of the matrix element.

Input for "PCIBAEM"

The program first requests the multipolarity of the transitions which are to be calculated. Next it asks for the parameters in the operator for the particular multipolarity you have selected. There is then the possibility of entering the initial- and final-state spin and parities for which the actual calculation should be performed. The transitions between all the states with a given spin and parity for which eigenvectors are available will always be calculated. If the spin of the initial state is entered as a negative number, matrix elements divided by $\sqrt{2j_i + 1}$ (i.e., the quantity of (5.22)), are given instead of the transition probability. Upon entering a value greater than 50 for the initial spin, the program will ask you to select another multipolarity. The defaults of the parameters are those entered last or otherwise those given in Table 5.2.

Table 5.2. Input parameters for "PCIBAEM", with their default values.

Variable Name	Default Value	Meaning
E0	1.	
M1,M1ND,M1Q	1.,0.,0.	
E2SD,E2DD	1.,CHQ	The default, CHQ, is the value used in the Hamiltonian.
E4DD	1.	
E1DF,E1Q	1.,0.	
E3SF,E3DF,E3Q	1.,0.,0.	

Example of an "FBEM" Run

Input:
(Same conventions as in Sect. 5.1.)
Transition Multipolarity (E0,M1,E1,E2,E3,E4,S=STOP)? E2 [R]
E2 Transitions
parameters: E2SD= 1.0000 , E2DD= −2.9580
are these ok? Y/N [R]
initial state spin = 2 [R]
initial state parity = 1 [R]
final state spin = 0 [R]
E2 transitions from 2+ to 0+
1=>1= 22.5160 /;/ 1=>2= 0.1962 /;/ 1=>3= 0.0031 /;/ 1=>4= 0.0000 /;/
2=>1= 0.0487 /;/ 2=>2= 13.8784 /;/ 2=>3= 0.8585 /;/ 2=>4= 0.0262 /;/
3=>1= 0.0972 /;/ 3=>2= 0.0406 /;/ 3=>3= 0.1738 /;/ 3=>4= 0.1347 /;/
4=>1= 0.0002 /;/ 4=>2= 0.0030 /;/ 4=>3= 7.2006 /;/ 4=>4= 0.3951 /;/
initial state spin = −2 [R]

initial state parity = 1 [R]
final state spin = 0 [R]
E2 transitions from 2+ to 2+
1=>1= −5.7031 /;/ 1=>2= 0.4833 /;/ 1=>3= 0.3236 /;/ 1=>4= −0.0705 /;/
2=>1= 0.4833 /;/ 2=>2= −3.8110 /;/ 2=>3= 2.4709 /;/ 2=>4= 0.8581 /;/
3=>1= 0.3236 /;/ 3=>2= 2.4709 /;/ 3=>3= 3.7915 /;/ 3=>4= −0.6983 /;/
4=>1= −0.0705 /;/ 4=>2= 0.8581 /;/ 4=>3= −0.6983 /;/ 4=>4= 2.3342 /;/
initial state spin = 100 [R]
Transition Multipolarity (E0,M1,E1,E2,E3,E4,S=STOP)? S [R]

Note that the responses involving letters, in this example "E2" and "S" have to be entered in upper case.

Output:
See file "PCIBAEM.OUT". Most of the output is self-explanatory. For the $2^+ \rightarrow 0^+$ transition B(E2) values have been calculated. For $2^+ \rightarrow 2^+$ transitions the matrix elements divided by a spin factor ($\sqrt{5}$ in this case) are printed. The test output accompanying the program was generated by this dialogue (assuming the same conventions as for the "PCIBAXW" example above). Note that the test problem for PCIBAXW must have been run previously so that the wave functions are available in "PHINT.WAV".

5.4.3   Program CFPGEN

In this program coefficients of fractional parentage (CFP) [5.12] are calculated for the *d*-bosons using analytical formulas, which are a simple extension of the ones given in Ref. [5.1].

The CFP are written to the file PHINT.CFP. This file must be created only once. In its default setup the CFP are calculated for a system with at most 7 bosons. Comment statements in the program indicate how this limit can be changed. Note that if this number is changed, the other programs and the library should be updated as well (i.e., changed, recompiled, and linked) according to the instructions given in the source code of CFPGEN. If this is not done correctly, the other programs will generate error messages, or, even worse, produce erroneous results.

This program does not require input.

5.5   How to Calculate for a Single Nucleus

A. Determine the number of bosons (=NPHMAX)
The number of bosons is equal to the total number of fermion pairs (particles or holes, whichever is less) outside a closed shell [5.6,7].

Example: $^{104}_{46}\mathrm{Pd}_{58}$

Neutrons: −4 fermions   +2 bosons

Protons: +8 fermions   +4 bosons
Total: 6 bosons.

B. Determine the "strategy": which limiting case?

If the spectrum is more rotational-like it is better to use the multipole representation for the Hamiltonian (5.6). This is to be preferred since it is possible to produce an exact rotational spectrum [5.2,3] using the parameters QQ and ELL only.

For a vibrational nucleus it is easier *not* to use the multipoles but to define the Hamiltonian in terms of HBAR, $C_{0,2,4}$, F, and G. In this case these parameters have a more direct interpretation [5.1]. It is important to note that these representations are equivalent and can produce the same results.

C. Fit the parameters in the Hamiltonian.

Make a first guess of the parameters, using the analytic formulas given in Sect. 5.6 for a nucleus closer to the SU(3) or O(6) limit. For a simple vibrational nucleus, the recipe for a first guess is

> HBAR $\approx E\,2^+$ and C=0., 0., 0.,
> F (normally positive) is large ($\approx +0.1$)
> if $B(E2; 2_2^+ \to 2_1^+) \approx \frac{1}{2}B(E2; 4_1^+ \to 2_1^+)$.
> G (always negative) is large (≈ -0.1)
> if $B(E2; 0_2^+ \to 2_1^+) \approx \frac{1}{2}B(E2; 4_1^+ \to 2_1^+)$.

If a neighboring nucleus has already been fitted, it is better to use these parameters.

Then, do a more detailed fit of the parameters in the Hamiltonian. Make several runs with slightly different values for the parameters in the Hamiltonian, and compare the calculated energies and $B(E2)$ values with experiment.

D. Calculate the $B(E2)$-Values.

Run the program PCIBAEM twice to calculate the matrix elements of the E2 operator (use negative initial spin) according to the following recipe. First set E2SD=1 and E2DD=0 and equate $M(E2; 2_1^+ \to 0_1^+) = A$ and $M(E2; 2_2^+ \to 0_1^+) = B$; then set E2SD=0 and E2DD=1 and equate $M(E2; 2_1^+ \to 0_1^+) = A'$ and $M(E2; 2_2^+ \to 0_1^+) = B'$. Here $M(E2)$ stands for the matrix elements; i.e., the values as printed by the program, multiplied by $\sqrt{2J+1}$, where J denotes the spin of the initial state. In this case, $J = 2$ for both the quantities A and B. Define the quantities $A_{\exp}$ and $B_{\exp}$ as the square roots of the experimental values for the $B(E2)$ multiplied by a similar factor $2J + 1$, giving in this case

$$A_{\exp} = \sqrt{5B(E2; 2_1^+ \to 0_1^+)} \quad \text{and} \quad B_{\exp} = \sqrt{5B(E2; 2_2^+ \to 0_1^+)}. \quad (5.23)$$

The values for E2SD and E2DD that reproduce these two transitions exactly can now be obtained by solving the following two linear equations:

$$E2SD \times A + E2DD \times A' = \pm A_{exp},$$
$$E2SD \times B + E2DD \times B' = \pm B_{exp}, \tag{5.24}$$

giving four solutions. The "physical" solution is restricted to E2SD> 0 and $-\sqrt{35}/2 \leq E2DD/E2SD < 0$

Instead of the $B(E2; 2_2^+ \to 0_1^+)$-value, as is taken in the above example, any other $B(E2)$ could have been taken, provided it is sensitive to the ratio E2DD/E2SD. The $B(E2; 4_1^+ \to 2_1^+)$, for example, is not suitable.

E. Negative parity states.
In principle the full interaction H_{df} (5.5) can be used, but usually a reasonable fit can be obtained using the multipole form (5.9). The energy of the f-boson EPS3 only has the effect of shifting all the negative-parity states by an equal amount and does not change the wave functions. E1 transitions can be fitted in a similar way, as was done for the E2 transitions.

5.6   Analytic Formulas

For some special forms of the Hamiltonian there exist analytic formulas for energies and transition matrix elements [5.1–3]. Only the formulas concerning energies and B(E2) values for positive parity states will be given here. More detailed information is given in Refs. [5.1–3].

5.6.1   The SU(5) Limit or the Anharmonic Vibrator

Hamiltonian [5.1] (see also (5.3))

$$H = \mathrm{HBAR}\,\hat{n}_d + \sum_{\substack{L=0,2,4 \\ \text{or } I=1,2,3}} C(I)\tfrac{1}{2}\sqrt{2L+1}[(d^\dagger d^\dagger)^{(L)} \times (dd)^{(L)}]^0 . \tag{5.25}$$

In the following we will use the conventional notation C_0, C_2, C_4 instead of $C(1)$, $C(2)$, and $C(3)$.

Analytic formula for the excitation energies of the Hamiltonian given in (5.25)

$$E([N], n_d, n_\beta, n_\delta, 1) = \mathrm{HBAR}\,n_d + \tfrac{1}{2}\alpha n_d(n_d - 1)$$
$$+ 2\beta n_\beta(2n_d - 2n_\beta + 3) + \gamma(L(L+1) - 6n_d) , \tag{5.26}$$

where

$$\alpha = (4C_2 + 3C_4)/7 , \tag{5.27}$$
$$\beta = (7C_0 - 10C_2 + 3C_4)/70 , \tag{5.28}$$
$$\gamma = -(C_2 - C_4)14 . \tag{5.29}$$

The quantum numbers n_d, n_β, n_δ, and L, all ≥ 0, should obey the following relations:

$$0 \leq n_d \leq N \quad \text{and} \quad 0 \leq M \leq L \leq 2M \tag{5.30}$$

with the exception that $L \neq 2M - 1$, where $M = n_d - 2n_\beta - 3n_\delta$.

Some B(E2) values
a. in the ground-state band:
$$B(E2; L + 2 \to L) = (\text{E2SD})^2(N - n_d)(n_d + 1), \tag{5.31}$$
where $N = n_s + n_d$ and $L = 2n_d$.
b. B(E2) leading to the 2_1^+-state:
$$\begin{aligned} B(E2; 4_1^+ \to 2_1^+) &= B(E2; 2^+2_2 \to 2_1^+) , \\ &= B(E2; 0_2^+ \to 2_1^+) , \\ &= \frac{2(N-1)}{N} B(E2; 2_1^+ \to 0_1^+) . \end{aligned} \tag{5.32}$$

The quadrupole moment of the 2^+-state

$$Q(2_1^+) = \text{E2DD} \frac{4}{5} \sqrt{\frac{2\pi}{7}}. \tag{5.33}$$

5.6.2   The SU(3) Limit or the Axial Rotor

Hamiltonian [5.2] (see also (5.6))

$$H = \tfrac{1}{2}\text{ELL}(\boldsymbol{L} \cdot \boldsymbol{L}) + \tfrac{1}{2}\text{QQ}(\boldsymbol{Q} \cdot \boldsymbol{Q}), \tag{5.34}$$

where $\boldsymbol{L} \cdot \boldsymbol{L}$ and $\boldsymbol{Q} \cdot \boldsymbol{Q}$ are defined in (5.7,8) The parameter CHQ should have its default value of $-\sqrt{35}/2$.

Analytic formula for the excitation energies of the Hamiltonian given in (5.34)

$$E([N], (\lambda, \mu), \kappa, L) = (\kappa' + \tfrac{3}{4}\kappa)L(L + 1) - \kappa C_2(\lambda, \mu), \tag{5.35}$$

where

$$\begin{aligned} C_2(\lambda, \mu) &= \lambda^2 + \mu^2 + \lambda\mu + 3(\lambda + \mu) , \\ \kappa &= -\text{QQ}/4 , \quad \text{and} \quad \kappa' = \text{ELL}/2 . \end{aligned} \tag{5.36}$$

The ground state band has the quantum numbers $(\lambda, \mu) = (2N, 0)$. The "β-" and "γ"-bands have $(\lambda, \mu) = (2N - 4, 2)$, giving for the excitation energy (E_x) of the 0_2^+-state

$$E_x([N], (2N - 4, 2), 0, 0) = 6\kappa(2N - 1) = -\frac{3}{2}\text{QQ}(2N - 1). \tag{5.37}$$

Some B(E2) values

In order that the E2 operator be the SU(3) quadrupole operator, its two parameters should have a definite ratio, E2DD/E2SD= $-\frac{1}{2}\sqrt{35}$ (the default value for E2DD). In the exact limit there are no E2 transitions between bands having different values for (λ, μ). The B(E2) values within the ground-state band $((\lambda, \mu) = (2N, 0))$ are given by

$$
\begin{aligned}
& B(E2; L+2 \rightarrow L) \\
& = (E2SD)^2 \frac{3(L+1)(L+2)}{4(2L+3)(2L+5)}(2N-L)(2N+L+3) .
\end{aligned} \tag{5.38}
$$

The quadrupole moment of the 2^+-state

$$
Q(2_1^+) = -E2SD\sqrt{\frac{2\pi}{5}}2(4N+3)/7. \tag{5.39}
$$

5.6.3 The O(6) Limit or the γ-Unstable Spectrum

Hamiltonian [5.3] (see also (5.6))

$$
H = \frac{1}{2}QQ(\boldsymbol{Q} \cdot \boldsymbol{Q}) + \frac{1}{2}ELL(\boldsymbol{L} \cdot \boldsymbol{L}) - \frac{5}{\sqrt{7}}OCT \left[(d^\dagger \tilde{d})^{(3)} \times (d^\dagger \tilde{d})^{(3)}\right]^{(0)} \tag{5.40}
$$

with CHQ=0.

Analytic formula for the excitation energies of the Hamiltonian given in (5.40)

$$
\begin{aligned}
& E([N], \sigma, \tau, n_\delta, L) \\
& = A(N-\sigma)(N+\sigma+4)/4 + B\tau(\tau+3)/6 + CL(L+1),
\end{aligned} \tag{5.41}
$$

where $A = -2\,QQ$, $B = 15\,OCT - 3\,QQ$, and $C = (ELL - OCT)/2$. The allowed values of σ, τ, n_δ, and L are given by

$$
\begin{aligned}
\sigma &= N, N-2, N-4, \ldots 0 \text{ or } 1, \\
\tau &= 0, 1, 2, \ldots \sigma, \\
n_\delta &= 0, 1, 2, \ldots, \text{ with } \lambda = \tau - 3n_\delta > 0 \\
L &= 2\lambda, 2\lambda-2, 2\lambda-3, \ldots \lambda.
\end{aligned}
$$

Some B(E2) values

In order to have the O(6) quadrupole operator one has to put E2DD = 0. In the exact limit only those B(E2) are allowed that satisfy the selection rules $\Delta\tau = \pm 1$ and $\Delta\sigma = 0$. The B(E2) in the ground-state band $(\sigma = N)$ are given by

$$
B(E2; L+2 \rightarrow L) = (E2SD)^2 \frac{L+2}{2(L+5)} \frac{(2N-1)(2N+L+8)}{4}. \tag{5.42}
$$

The quadrupole moment of the 2^+-state. The quadrupole moments of all states are equal to zero in the exact limit, in particular

$$
Q(2_1^+) = 0. \tag{5.43}
$$

5.7   Formulas Used in the Programs

5.7.1   Multipole Expansion (the Relation between (5.3) and (5.6))

$$\begin{aligned}
\text{HBAR} &= \text{EPS} + \tfrac{1}{2}(\chi^2 - 4)\text{QQ} + 3\,\text{ELL} + 7\,\text{OCT} + 9\,\text{HEX}\,, \\
C(1) &= -6\,\text{ELL} + \chi^2\text{QQ} - 14\,\text{OCT} + 18\,\text{HEX}\,, \\
C(2) &= -3\,\text{ELL} - \frac{3}{14}\chi^2\text{QQ} + 9\,\text{OCT} + \frac{36}{7}\text{HEX}\,, \\
C(3) &= 4\,\text{ELL} + \frac{2}{7}\chi^2\text{QQ} + \text{OCT} + \frac{1}{7}\text{HEX}\,, \\
F &= \sqrt{5}\chi\text{QQ}\,, \\
G &= \tfrac{1}{2}\sqrt{5}\text{QQ}\,, \\
\text{CH1} &= 0\,, \\
\text{CH2} &= \text{QQ}\,, \\
\chi &= \text{CHQ}/\sqrt{5}\,.
\end{aligned}$$

In calculating excitation energies, only six parameters are linearly dependent.

The contribution of CH1 and CH2 can be absorbed in HBAR, $C(I)$, I=1,2,3 and a constant term, which only contributes to binding energies,

$$\text{HBAR}' = \text{HBAR} + (N-1)(\text{CH2} - \text{CH1})\,,$$

$$C'(I) = C(I) + \text{CH1} - 2\,\text{CH2}\,.$$

Here, N is the total number of bosons.

5.7.2   Projection of IBA-2 Parameters onto those of IBA-1 (Relation between (5.3) and (5.13))

$$\begin{aligned}
\text{HBAR} &= \text{ED}\,, \\
C(I) &= \frac{N_\nu(N_\nu - 1)}{N(N-1)}\text{CLN}(I) + \frac{N_\pi(N_\pi - 1)}{N(N-1)}\text{CLP}(I) \\
&\quad + 2\kappa\frac{N_\nu N_\pi}{N(N-1)}\left(-1 - \frac{\chi_\nu + \chi_\pi}{\sqrt{7}} + \alpha_I\chi_\nu\chi_\pi\right)\,, \\
\alpha_1 &= 1\,, \\
\alpha_2 &= -\frac{3}{14}\,, \\
\alpha_3 &= \frac{2}{7}\,, \\
F &= \kappa(\chi_\nu + \chi_\pi)\frac{N_\nu N_\pi}{N(N-1)}\sqrt{5}\,,
\end{aligned}$$

$$G \quad = \frac{N_\nu N_\pi}{N(N-1)}\sqrt{5}\,,$$

$$\mathrm{CH1} \quad = -2\kappa\frac{N_\nu N_\pi}{N(N-1)}\left(1+\frac{\chi_\nu+\chi_\pi}{\sqrt{7}}\right)\,,$$

$$\mathrm{CH2} \quad = -2\kappa\frac{N_\nu N_\pi}{N(N-1)}\frac{\chi_\nu+\chi_\pi}{\sqrt{7}}\,,$$

where N_ν = NN, N_π = NP, κ = RKAP, χ_ν = CHN, χ_π = CHP, $N = N_\pi + N_\nu$.

5.8   Technical Notes

The three main programs mentioned above partially require the same subroutines. For this reason the FORTRAN source is split into several files which must be combined to yield the executables. The source files are

PCIBAXW.FOR Main program and some subroutines for calculating excitation energies and wave functions.

PCIBAEM.FOR Main program and some subroutines for calculating electromagnetic transition matrix elements and probabilities.

CFPGEN.FOR Main program and some subroutines for calculating coefficients of fractional parentage.

PCIBALIB.FOR Additional subroutines used by both PCIBAXW and PCIBAEM (preferably used as a library).

ANGMOM.FOR Routines for calculating angular-momentum recoupling brackets (preferably used as a library).

DIAG.FOR Routines for the diagonalization of a real symmetric matrix.

The FORTRAN files should be individually compiled and then linked to form the executable programs as follows:

PCIBAXW: needs PCIBAXW, DIAG, and some subroutines of PCIBALIB and ANGMOM.
PCIBAEM: needs PCIBAEM, and some subroutines of PCIBALIB and ANGMOM.
CFPGEN: needs CFPGEN and ANGMOM.

At this stage, CFPGEN should be run first to generate the binary file PHINT.CFP, which contains all required values of the Racah coefficients. For applications, PCIBAXW should then be run to generate the spectra and wave functions, followed by PCIBAEM for the transition rates.

   Sample outputs are in the accompanying files "PCIBAXW.OUT" and "PCIBAEM.OUT". They correspond to the output generated for the sample input given in Sect. 5.4.

   If the program is run with debugging enabled, error messages concerning the overflow of some indices for the use of various matrices and vectors in the

codes may be generated. These should be ignored; they are simply caused by
the widespread practice of not giving the dimensions correctly for subroutine
arguments.

References

[5.1]   A. Arima and F. Iachello, Ann. Phys. **99** (1976) 253

[5.2]   A. Arima and F. Iachello, Ann. Phys. **111** (1978) 201

[5.3]   A. Arima and F. Iachello, Ann. Phys. **123** (1979) 468

[5.4]   Further reading: R.F. Casten and D.D. Warner, Rev. Mod. Phys. **60** (1988)
         389; F. Iachello and A. Arima, *The interacting boson model* (Cambridge
         University Press, 1987); *Interacting Bosons in Nuclear Physics*, edited by
         F. Iachello (Plenum, New York, 1979); *Interacting Bose-Fermi Systems in
         Nuclei*, edited by F. Iachello, (Plenum, New York, 1981)

[5.5]   A. Arima and F. Iachello, Phys. Rev. **C16** (1977) 2085

[5.6]   A. Arima, T. Otsuka, F. Iachello, and I. Talmi, Phys. Lett. **76B** (1978)
         139.

[5.7]   T. Otsuka, A. Arima, and F. Iachello, Nucl. Phys. **A309** (1978) 1; T.
         Otsuka, Ph.D. Thesis, University of Tokyo, 1979; A. Arima, T. Otsuka, F.
         Iachello and I. Talmi, Phys. Lett. **66B** (1977) 205

[5.8]   O. Scholten, Ph.D. Thesis, University of Groningen, 1980; O. Scholten in
         Progress in Particle and Nuclear Physics, edited by A. Faessler (Pergamon,
         Oxford, 1985) vol. 14, p. 189

[5.9]   O. Scholten, F. Iachello, and A. Arima, Ann. Phys. **115** (1979) 325

[5.10]   J.N. Ginoccio and M.W. Kirson, Phys. Lett **44** (1980) 1744; A.E.L.
         Dieperink and O. Scholten, Nucl. Phys. **A346** (1980) 125

[5.11]   B. Eckhardt, Phys. Rep. **163** (1987) 205

[5.12]   B. Bayman and A. Landé, Nucl. Phys. **77** (1966) 1

[5.13]   T. Otsuka and O. Scholten, program NPBOS, KVI Int. Report 253 (1980)

6. Numerical Application of the Geometric Collective Model

D. Troltenier, J.A. Maruhn, and P.O. Hess

6.1 Introduction

The generalized collective model (GCM) is a phenomenological model for the description of the low-energy ($< 3\,\mathrm{MeV}$) collective properties of even–even nuclei; i.e., nuclei with even charge and neutron number. The physical picture behind the model is that these nuclei behave like incompressible liquid drops, especially for higher mass numbers. Thus, the model neglects single particle properties and determines the physical properties of the nucleus by its shape. From this point of view it is clear that the excitations of the nucleus within this model are vibrations and rotations. The theoretical problem is the classification of the excited states by quantum numbers. This is possible by means of group-theoretical considerations. For an extended discussion, we refer to Refs. [6.1–6].

The computer code presented in this chapter allows the computation of the energy spectrum, transition probabilities, etc. of a given collective Hamiltonian. In addition it can determine a set of parameters for the Hamiltonian such that the experimental data for a certain nucleus are fitted in a least-squares sense.

6.2 Collective Parameters

The shape of a nucleus can be described by an expansion in spherical harmonics:

$$R(\vartheta, \varphi, t) = R_0 \left(1 + \sum_{\lambda,\mu} \alpha_{\lambda\mu}^*(t) Y_{\lambda\mu}(\vartheta, \varphi) \right) , \tag{6.1}$$

where $R(\vartheta, \varphi, t)$ is the radius of the nucleus along the direction (ϑ, φ) in the (laboratory-fixed) center-of-mass system, and $\alpha_{\lambda\mu}$ are the multipole deformation parameters [6.1,7]. For low-energy spectra by far the most important mode is the quadrupole. The present version of the model thus restricts the expansion to

$$R(\vartheta, \varphi, t) = R_0 \left(1 + \alpha_{00}(\alpha_{2\mu}, t) Y_{00} + \sum_{\mu} \alpha_{2\mu}^*(t) Y_{2\mu}(\vartheta, \varphi) \right) . \tag{6.2}$$

The term involving α_{00} in the parentheses is not independent, but merely serves to ensure volume conservation. Calculating the nuclear volume to second order yields the condition

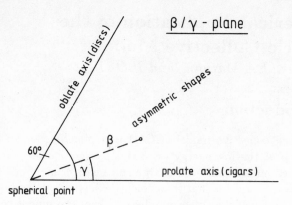

Fig. 6.1. The β–γ-plane with the different types of nuclear shape indicated.

$$\alpha_{00} = -\frac{1}{\sqrt{4\pi}} \sum_\mu \alpha_{2\mu}\alpha_{2\mu}^*. \tag{6.3}$$

It is useful to study the nucleus in the body-fixed principal-axis system, the "intrinsic" system. The transformation of the $\alpha_{2\mu}$ to the deformation variables in the intrinsic system a_{2m} is given by a rotation:

$$\alpha_{2\mu} = \sum_\nu D_{\mu\nu}^{2^*}(\vartheta_1,\vartheta_2,\vartheta_3)\,a_{2\nu}, \tag{6.4}$$

with $D_{\mu\nu}^{\lambda^*}(\vartheta_1,\vartheta_2,\vartheta_3)$ the well-known Wigner functions depending on the Euler angles $\vartheta_1,\vartheta_2,\vartheta_3$, which determine the orientation of the principal axes of the nucleus in the laboratory frame. Since the off-diagonal moments vanish in the intrinsic system, the following equations hold:

$$a_{2-2} = a_{22},$$
$$a_{2-1} = a_{21} = 0.$$

These coordinates are customarily expressed in terms of polar coordinates β and γ:

$$a_{20} = \beta\cos\gamma,$$
$$a_{22} = \frac{1}{\sqrt{2}}\beta\sin\gamma. \tag{6.5}$$

The physical meaning of these variables is illustrated in Fig. 6.1.

6.3 The Hamiltonian

In the laboratory system the Hamiltonian is given as a sum of kinetic and potential energy. All terms must be rotationally invariant, and this may be achieved either by coupling the spherical tensors to total angular momentum zero, or by demanding independence of the nuclear orientation as embodied in the Euler angles. The former method is preferable in the laboratory system, while the latter allows one to construct the potential, e. g., as a function of β and γ alone [6.7,8]. We will use whichever of these is more convenient for a given problem. The Hamiltonian has the general form

$$\hat{H} = \hat{T} + V. \tag{6.6}$$

Ignoring the problems of quantization [6.1], we assume the definition of momentum operators $\hat{\pi}_{2\mu}$ fulfilling the usual canonical commutation relations with the deformation coordinates:

$$[\hat{\pi}_{2\mu}, \alpha_{2\nu}] = -i\hbar\delta_{\mu\nu}. \tag{6.7}$$

The kinetic energy is constructed to contain the two lowest-order terms proportional to the square of the momenta:

$$\hat{T} = \frac{1}{B_2}[\hat{\pi} \times \hat{\pi}]^{[0]} + \frac{P_3}{3}\left\{\left[[\hat{\pi} \times \alpha]^{[2]} \times \hat{\pi}\right]^{[0]}\right\}, \tag{6.8}$$

where $\{\dots\}$ means the sum over all permutations.

For the potential we use a polynomial expansion in the deformation variables, containing all possible independent terms up to sixth order:

$$
\begin{aligned}
V(\beta,\gamma) &= C_2\left[\alpha \times \alpha\right]^{[0]} + C_3\left[[\alpha \times \alpha]^{[2]} \times \alpha\right]^{[0]} \\
&\quad + C_4\left([\alpha \times \alpha]^{[0]}\right)^2 + C_5\left[[\alpha \times \alpha]^{[2]} \times \alpha\right]^{[0]}[\alpha \times \alpha]^{[0]} \\
&\quad + C_6\left([[\alpha \times \alpha]^{[2]} \times \alpha]^{[0]}\right)^2 + D_6\left([\alpha \times \alpha]^{[0]}\right)^3 \\
&= C_2\frac{1}{\sqrt{5}}\beta^2 - C_3\sqrt{\frac{2}{35}}\beta^3\cos 3\gamma + C_4\frac{1}{5}\beta^4 \\
&\quad - C_5\sqrt{\frac{2}{175}}\beta^5\cos 3\gamma + C_6\frac{2}{35}\beta^6\cos^2 3\gamma + D_6\frac{1}{5\sqrt{5}}\beta^6. \tag{6.9}
\end{aligned}
$$

The second version corresponds to the intrinsic system.

6.4   The Wave Functions

The eigenstates of the above Hamiltonian are calculated by diagonalization in a basis. We choose as basic functions the eigenfunctions of the five-dimensional harmonic oscillator $\hat{H}_5$:

$$\hat{H}_5 = \frac{\sqrt{5}}{2B_2'}[\hat{\pi} \times \hat{\pi}]^{[0]} + \frac{\sqrt{5}}{2}C_2'[\alpha \times \alpha]^{[0]}, \tag{6.10}$$

with B_2', C_2' the usual mass and stiffness parameters. Transformed to the principal-axis system and expressed in the intrinsic variables $\hat{H}_5$ reads

$$\hat{H}_5 = -\frac{\hbar^2}{2B_2'}\left(\frac{1}{\beta^4}\frac{\partial}{\partial\beta}\beta^4\frac{\partial}{\partial\beta} - \frac{1}{\beta^2}\hat{\Lambda}^2\right) + \frac{C_2'}{2}\beta^2, \tag{6.11}$$

with $\hat{\Lambda}^2$ the *seniority* operator:

$$\hat{\Lambda}^2 = -\frac{1}{\sin 3\gamma}\frac{\partial}{\partial\gamma}\sin 3\gamma\frac{\partial}{\partial\gamma} + \sum_{k=1}^{3} I_k^{-1}\hat{L}_k'^2\left(\vartheta_1, \vartheta_2, \vartheta_3\right). \tag{6.12}$$

I_k is the moment of inertia with respect to the intrinsic axis k $(k = 1, 2, 3)$:

$$I_k = 4B_2'\sin^2\left(\gamma - \frac{2\pi}{3}k\right), \tag{6.13}$$

and $\hat{L}_k'\left(\vartheta_1, \vartheta_2, \vartheta_3\right)$ are the intrinsic components of the angular-momentum operator.

The construction of the eigenfunctions of the five-dimensional harmonic oscillator proceeds most smoothly if group-theoretical methods are used. We choose the following group chain for the complete classification of the states:

$$U(5) \supset O(5) \supset O(3) \supset O(2). \tag{6.14}$$

This group chain has the advantage of including the physically important angular-momentum quantum numbers L and M as eigenvalues of the Casimir operators of $O(3)$ and $O(2)$. It starts with the symmetry group of the five-dimensional harmonic oscillator, $U(5)$, which has the phonon-number operator $\hat{N}$ with eigenvalue ν as its Casimir operator:

$$\hbar\omega_2\hat{N} = \hat{H}_5 - \frac{5}{2}\hbar\omega_2 = \hbar\omega_2\sum_{m=-2}^{2}\hat{\beta}_m^+\hat{\beta}_m, \tag{6.15}$$

$\hat{\beta}_m^+$ and $\hat{\beta}_m$ being the usual creation and annihilation operators for quadrupole phonons

$$\hat{\beta}_{2\mu}^+ = \frac{1}{2}\left[\sqrt{\frac{2}{\hbar}}\sqrt{\frac{B_2'}{C_2'}}\alpha_{2\mu} - i\sqrt{\frac{2}{\hbar}}\frac{1}{\sqrt{B_2'C_2'}}\hat{\pi}_{2\mu}^+\right], \tag{6.16}$$

$$\hat{\beta}_{2\mu} = \frac{1}{2}\left[\sqrt{\frac{2}{\hbar}}\sqrt{\frac{B_2'}{C_2'}}\alpha_{2\mu}^+ + i\sqrt{\frac{2}{\hbar}}\frac{1}{\sqrt{B_2'C_2'}}\hat{\pi}_{2\mu}\right]. \tag{6.17}$$

Using these operators all possible states may now be constructed, taking into account angular-momentum coupling and boson symmetry. We skip this lengthy construction, which may be found in [6.5,6], and only present the crucial results.

The quantum numbers obtained from the group chain are:

Group	U(5)	O(5)	O(3)	O(2)
Casimir Operator	$\hat{N}$	$\hat{\Lambda}^2$	$\hat{L}^2$	$\hat{L}_z$
Quantum Number	ν	λ	L	M

and the eigenstate is denoted by $|\nu\lambda\mu LM\rangle$. The physical meaning of the quantum numbers is as follows: ν is the number of phonons, λ denotes the number of phonons that are not coupled pairwise to zero and L and M are the angular momentum and its projection in the laboratory system. Unfortunately these quantum numbers are not sufficient to classify all states uniquely owing to a degeneracy for $L \geq 6$. The quantum number μ counts simply these degenerated states. Physically it has to be identified with the number of phonon triplets coupled to angular momentum zero.

The quantum numbers can only assume positive integer values and have to obey the selection rules [6.3]

$$\frac{L}{2} \leq \lambda - 3\mu \leq L, \qquad L \text{ even} ;$$

$$\frac{L-3}{2} \leq (\lambda - 3) - 3\mu \leq L - 3, \qquad L \text{ odd} . \tag{6.18}$$

Let us give some examples for the selection rules. The simplest nontrivial state is of course $|1002M\rangle$, with only one phonon and angular momentum two; i.e., a quadrupole phonon. Consider $L = 4$, $\lambda = 4$. In this case we must have $0 \leq \mu \leq 2/3$. Since μ is an integer, only $\mu = 0$ is allowed. This is easily understood, since for say $\mu = 1$ three phonons would couple to angular momentum zero, and the last phonon should couple to angular momentum four, which is impossible. As a last example, take $L = 6$, $\lambda = 3$, which allows only $\mu = 0$, while for $L = 6$, $\lambda = 6$ we have both $\mu = 0$ and $\mu = 1$, since there may be a phonon triplet coupled to angular momentum zero with the other three phonons coupled to angular momentum six. In general the maximum value of μ for fixed λ and L is given by $INT(L/6)$.

The representation of the wave functions in coordinate space is given by:

$$|\nu\lambda\mu LM\rangle$$

$$= F_l^\lambda(\beta) \sum_K \Phi_K^{\lambda\mu L}(\gamma) \left[D_{MK}^{L^*}(\vartheta_1, \vartheta_2, \vartheta_3) + (-)^L D_{M-K}^{L^*}(\vartheta_1, \vartheta_2, \vartheta_3) \right] (6.19)$$

with $l = \frac{1}{2}(\nu - \lambda)$ and K even. In constructing the wave function it is also important to consider the ambiguities in the choice of the intrinsic coordinate system; this leads, for example, to the combination of rotation matrices in the parentheses (see Ref. [6.1]).

The solutions for the β-dependent part are

$$F_l^\lambda(\beta)$$

$$= \left[\frac{2(l!)}{\Gamma(l+\lambda+\frac{5}{2})} \right]^{\frac{1}{2}} \left(\frac{C_2'}{\hbar\omega_2} \right)^{\frac{5}{4}+\frac{\lambda}{2}} \beta^\lambda L_l^{\lambda+\frac{3}{2}} \left(\left(\frac{C_2'}{\hbar\omega_2} \right) \beta^2 \right) \exp\left(-\frac{1}{2} \frac{C_2'}{\hbar\omega_2} \beta^2 \right),$$

$$\tag{6.20}$$

with $\omega_2 = \sqrt{C_2'/B_2'}$ and $L_l^{\lambda+\frac{3}{2}}$ the Laguerre polynomials.

The analytical expressions for the γ-dependent part are much more complicated and too lengthy to be presented here. In any case, the code does not use them, as the integration of the γ-dependent matrix elements is done numerically. We only give their symmetry properties

$$\begin{aligned}
\Phi_K^{\lambda\mu L}(\gamma) &= (-1)^L \Phi_{-K}^{\lambda\mu L}(\gamma), \\
\hat{\Lambda}^2 \Phi_K^{\lambda\mu L}(\gamma) &= \lambda(\lambda+3)\hat{\Lambda}^2 \Phi_K^{\lambda\mu L}(\gamma)
\end{aligned}$$

and refer the interested reader to Ref. [6.4].

6.5 Matrix Elements Containing $(\alpha_{2\mu})^\rho$

With the definition

$$T_M^{\lambda\mu L}(\gamma)$$

$$= \sum_K \Phi_K^{\lambda\mu L}(\gamma) \left[D_{MK}^{L*}(\vartheta_1,\vartheta_2,\vartheta_3) + (-)^L D_{M-K}^{L*}(\vartheta_1,\vartheta_2,\vartheta_3) \right], \qquad (6.21)$$

the general form of a matrix element in the GCM is given by:

$$\langle \nu''\lambda''\mu''L''M'' | \beta^\rho T_K^{\lambda\mu L}(\gamma) | \nu'\lambda'\mu'L'M' \rangle$$

$$= 8\pi^2 (-1)^{L''-M''} \begin{pmatrix} L & L' & L'' \\ M & M' & M'' \end{pmatrix} \left(\int_0^\infty F_{n''}^{\lambda''}(\beta)\beta^\rho F_{n'}^{\lambda'}(\beta)\beta^4 \mathrm{d}\beta \right)$$

$$\times \int_0^\infty \sum_{KK'K''} \begin{pmatrix} L & L' & L'' \\ K & K' & K'' \end{pmatrix} \Phi_K^{\lambda\mu L}(\gamma)\Phi_{K'}^{\lambda'\mu'L'}(\gamma)\Phi_{K''}^{\lambda''\mu''L''}(\gamma) \sin 3\gamma \mathrm{d}\gamma.$$

$$(6.22)$$

In the last equation the integration over the Euler angles has already been performed. Since the corresponding integral contains only Wigner functions, it is easily evaluated by using standard formulae [6.9].

It should be noted that the wave functions are not yet normalized, so that the above expressions have to be divided by

$$|\langle \nu''\lambda''\mu''L''M'' | \nu''\lambda''\mu''L''M'' \rangle|^{\frac{1}{2}}$$

and

$$|\langle \nu'\lambda'\mu'L'M' | \nu'\lambda'\mu'L'M' \rangle|^{\frac{1}{2}},$$

respectively. The integration over β yields [6.5]

$$\int_0^\infty F_{l'}^{\lambda'}(\beta)\beta^\rho F_l^{\lambda}(\beta)\beta^4 \mathrm{d}\beta$$

$$
= \left(\frac{\hbar\omega_2}{C_2'}\right)^{\frac{\ell}{2}} \left[\frac{l!}{\Gamma(l+\lambda+\frac{5}{2})}\right]^{\frac{1}{2}} \left[\frac{l'!}{\Gamma(l'+\lambda'+\frac{5}{2})}\right]^{\frac{1}{2}}
$$

$$
\times \Gamma\left(\tfrac{1}{2}(\rho+\lambda+\lambda'+5)\right)(-1)^{l+l''}\sum_\sigma(-1)^\sigma \begin{pmatrix}\tfrac{1}{2}(\rho+\lambda'-\lambda)\\ l-\sigma\end{pmatrix}
$$

$$
\times \begin{pmatrix}\tfrac{1}{2}(\rho+\lambda-\lambda')\\ l'-\sigma\end{pmatrix}\begin{pmatrix}-\tfrac{1}{2}(\rho+\lambda+\lambda'+5)\\ \sigma\end{pmatrix}. \tag{6.23}
$$

Calculating these formulae explicitly produces the formulae used in subroutine RADME (for **rad**ial **m**atrix **e**lements). Two cases have to be distinguished, $\lambda' - \lambda \le \rho$ and $\lambda' - \lambda \ge \rho$. For the first case the following formula for the matrix elements of β^ρ is obtained:

$$
\int_0^\infty F_{l'}^{\lambda'}(\beta)\beta^\rho F_l^\lambda(\beta)\beta^4 \mathrm{d}\beta
$$

$$
= \left(\frac{\hbar\omega_2}{C_2'}\right)^{\frac{\ell}{2}} \left[\frac{l!}{\Gamma(l+\lambda+\frac{5}{2})}\right]^{\frac{1}{2}} \left[\frac{l'!}{\Gamma(l'+\lambda'+\frac{5}{2})}\right]^{\frac{1}{2}} \Gamma\left(\tfrac{1}{2}(\rho+\lambda'-\lambda+2)\right)
$$

$$
\times \Gamma\left(\tfrac{1}{2}(\rho+\lambda-\lambda'+2)\right)(-1)^{l+l''}
$$

$$
\times \sum_\sigma \frac{\Gamma\left(\tfrac{1}{2}(\rho+\lambda+\lambda'+5)+\sigma\right)}{\sigma!(l-\sigma)!(l'-\sigma)!\Gamma\left(\sigma+\tfrac{1}{2}(\rho+\lambda'-\lambda)-l+1\right)}
$$

$$
\times \frac{1}{\Gamma\left(\sigma+\tfrac{1}{2}(\rho+\lambda-\lambda')-l'+1\right)}.
$$

The bounds of the summation are $0 \le \sigma \le \min(l, l')$. For the second case we can choose $\lambda' > \lambda$ without any restriction to get

$$
\int_0^\infty F_{l'}^{\lambda'}(\beta)\beta^\rho F_l^\lambda(\beta)\beta^4 \mathrm{d}\beta
$$

$$
= \left(\frac{\hbar\omega_2}{C_2'}\right)^{\frac{\ell}{2}} \left[\frac{l!}{\Gamma(l+\lambda+\frac{5}{2})}\right]^{\frac{1}{2}} \left[\frac{l'!}{\Gamma(l'+\lambda'+\frac{5}{2})}\right]^{\frac{1}{2}}
$$

$$
\times (-1)^l \frac{\Gamma\left(\tfrac{1}{2}(\rho+\lambda'-\lambda+2)\right)}{\Gamma\left(\tfrac{1}{2}(-\rho+\lambda'-\lambda)\right)}
$$

$$
\times \sum_\sigma (-1)^\sigma \frac{\Gamma\left(\tfrac{1}{2}(\rho+\lambda+\lambda'+5)+\sigma\right)\Gamma\left(\tfrac{1}{2}(\lambda'-\lambda-\rho)+l'-\sigma\right)}{\sigma!(l-\sigma)!(l'-\sigma)!\Gamma\left(\sigma+\tfrac{1}{2}(\rho+\lambda'-\lambda)-l'+1\right)}. \tag{6.24}
$$

There are two different cases for the summation bounds:

$$
\max(l'-\tfrac{1}{2}(\rho+\lambda'-\lambda),0) \le \sigma \le \min(l,l'),\ \text{for } \tfrac{1}{2}(\rho+\lambda'-\lambda)\ \text{even}
$$

and
$$0 \leq \sigma \leq \min(l, l'), \text{ for } \tfrac{1}{2}(\rho + \lambda' - \lambda) \text{ odd,}$$
respectively.

For the γ-dependent matrix elements a numerical evaluation is much more convenient and even more economical than straightforward use of the analytical results. We therefore provide files containing the values of these matrix elements; the file names are `ANGP.DAT` for those of the Hamiltonian and `ANGQ.DAT` for the matrix elements of the quadrupole operator. These matrix elements include the integration over the Euler angles and the proper normalization factors.

The files contain only the nonzero matrix elements, three on each line, each preceeded by the indices of its left and right basis functions. A block of matrix elements is delimited by a line that contains fewer than three matrix elements (so that the formatted input will yield values of zero, which can be checked by the code). The blocks in `ANGP.DAT` contain the matrix elements of the individual parts of the Hamiltonian, while `ANGQ.DAT` contains alternating blocks of matrix elements of α and α^2 for various angular-momentum combinations. These combinations are $(0,2)$, $(2,2)$, $(2,3)$, $(2,4)$, $(3,4)$, $(4,4)$, $(3,5) \cdots (I-2,I)$, $(I-1,I)$, (I,I), $(I-1,I+1)$, $\cdots$, $(10,10)$. The matrix elements of $(3,3)$ vanish identically. This means e.g. that $L = 3$-states have no quadrupole moment.

The matrix elements of the potential are ordered in an analogous way. But since the potential has angular momentum zero only matrix elements between states of the same angular momentum appear. Again we have different operators $\cos(3\gamma)^{\mu}$, $(\mu = 0, 1, 2)$ for which the matrix elements are ordered in blocks of angular momenta L, $(L = 0, 2, 3, \ldots, 10)$.

The selection rules for the matrix elements of (6.22) can now be summarized as

$$|\lambda'' - \lambda'| \leq \lambda,$$
λ even $\Leftrightarrow \Delta\lambda$ even $\Leftrightarrow \Delta\nu$ even,
λ odd $\Leftrightarrow \Delta\lambda$ odd $\Leftrightarrow \Delta\nu$ odd,
$\Delta\lambda \leq \rho$, $\tfrac{1}{2}(\rho + \lambda' - \lambda)$ even $\Rightarrow \Delta\nu \leq \rho$.

For $L = 0$ we have a special case:

$$|\lambda'' - \lambda'| \leq 3\mu,$$
μ even $\Leftrightarrow \Delta\lambda$ even $\Leftrightarrow \Delta\nu$ even,
μ odd $\Leftrightarrow \Delta\lambda$ odd $\Leftrightarrow \Delta\nu$ odd.

6.6   Matrix Elements Containing $(\pi_{2\mu})^{\rho}$

For the calculation of the matrix elements of the kinetic energy the gradient formula [6.1,2,4] is employed. Consider the commutation relation between the momentum $\hat{\pi}_{2\mu}$ and $\hat{H}_5$:

$$\left[\hat{\pi}_{2\mu}, \hat{H}_5\right] = -i\hbar C_2' \alpha_{2\mu}^* \tag{6.25}$$

Taking the matrix element of this equation yields

$$\langle \nu'\lambda'\mu'L'M'|\hat{\pi}_{2\mu}|\nu\lambda\mu LM\rangle = i\frac{C_2'}{\omega_2}\frac{\langle \nu\lambda\mu LM|\hat{\alpha}_{2\mu}|\nu'\lambda'\mu'L'M'\rangle}{\nu'-\nu} \tag{6.26}$$

With this formula we can calculate the matrix elements of the two π-dependent parts in the Hamiltonian:

$$\langle \nu'\lambda'\mu'L'M'|\left[\hat{\pi}\times\hat{\pi}\right]^0|\nu\lambda\mu LM\rangle$$
$$= (-1)^{\frac{\nu'-\nu}{2}}\left(\frac{C_2'}{\omega_2'}\right)^2 \langle \nu\lambda\mu LM|\left[\hat{\alpha}\times\hat{\alpha}\right]^0|\nu'\lambda'\mu'L'M'\rangle$$

and

$$\langle \nu'\lambda'\mu'L'M'|\left\{\left[[\hat{\pi}\times\alpha]\times\hat{\pi}\right]^0\right\}|\nu\lambda\mu LM\rangle$$
$$= F(-1)^{\frac{\nu'-\nu}{2}}\left(\frac{C_2'}{\omega_2'}\right)^2 \langle \nu\lambda\mu LM|\left[[\alpha\times\alpha]^2\times\alpha\right]^0|\nu'\lambda'\mu'L'M'\rangle$$

with $F = -3$ if $\nu'-\nu = \pm 3$ and $F = 1$ if $\nu'-\nu = \pm 1$. For all other differences in ν the above matrix element vanishes, as is easily seen observing that $\alpha_{2\mu}$ is essentially a sum of one creation and one annihilation operator.

6.7   The Quadrupole Operator $\hat{Q}_{2\mu}$

The quadrupole operator is the crucial quantity needed for the calculation of quadrupole moments and transition probabilities for electric quadrupole radiation. Neglecting possible spin contributions the microscopic operator is

$$\hat{Q}_{2\mu} = \sum_{i=1}^{Z} r_i^2 Y_{2\mu}^*(\vartheta,\varphi)\,. \tag{6.27}$$

The sum goes over all Z protons. Under the assumption of a homogenous charge distribution inside the radius $R(\vartheta,\varphi)$ the natural translation into the collective model is

$$\hat{Q}_{2\mu} = \frac{3Z}{4\pi R_0^3}\int\frac{R(\vartheta,\varphi)^5}{5}Y_{2\mu}^*(\vartheta,\varphi)d\Omega. \tag{6.28}$$

Inserting the radius yields, up to second order,

$$\hat{Q}_{2\mu} = \frac{3ZR_0^2}{4\pi}\left(\alpha_{2\mu}^* - \frac{10}{\sqrt{70\pi}}[\alpha\times\alpha]_{2\mu}^*\right). \tag{6.29}$$

Finally we have to transform the quadrupole operator into the intrinsic system according to (6.4). There is no need to write down the complete expression; just note that matrix elements of the following two quantities have to be evaluated:

$$\alpha_{2m} = \beta \left[\frac{1}{\sqrt{2}} \left(D^{2*}_{m2} + D^{2*}_{m-2} \right) \sin\gamma + D^{2*}_{m0} \cos\gamma \right],$$

$$[\alpha \times \alpha]_{2m} = \frac{\beta^2}{\sqrt{7}} \left[\left(D^{2*}_{m2} + D^{2*}_{m-2} \right) \sin(2\gamma) - \sqrt{2} D^{2*}_{m0} \cos(2\gamma) \right].$$

(For the sake of clarity the arguments of the Wigner functions are omitted.) The matrix elements of the quadrupole operator fulfill the following selection rules:

$$\alpha_{2\mu} \quad : \quad |\Delta\lambda| = 1, \quad |\Delta\nu| = 1 \quad |\Delta L| \le 2 \, ;$$

$$[\alpha \times \alpha]^2_m \quad : \quad |\Delta\lambda| = 0,2 \quad |\Delta\nu| = 0,2 \quad |\Delta L| \le 2 \, .$$

Physically relevant data connected with this operator are the reduced transition probability $B(E2, \delta_i L_i \to \delta_f L_f)$ and the quadrupole moment $Q(\delta, L)$. If $|\delta, LM\rangle$ is a collective state with angular momentum L, projection M, and all other quantum numbers summarized in δ, the $B(E2)$-value for the transition between the initial and the final state reads

$$B(E2, \delta_i L_i \to \delta_f L_f) = \frac{1}{2L_i + 1} \sum_{M_i, M_f, m} |\langle \delta_f, L_f M_f | \hat{Q}_{2m} | \delta_i, L_i M_i \rangle|^2$$

$$= \frac{2L_f + 1}{2L_i + 1} |\langle \delta_f L_f || Q_2 || \delta_i L_i \rangle|^2 \, , \tag{6.30}$$

where $|\langle \delta_f L_f || Q_2 || \delta_i L_i \rangle|^2$ is the reduced matrix element.

Since the code calculates only the upwards transition probabilities, we give the formula connecting upward and downward $B(E2)$s:

$$B(E2, L_f \to L_i) = \frac{2L_i + 1}{2L_f + 1} B(E2, L_i \to L_f) \, . \tag{6.31}$$

The formula for the quadrupole moment is given by

$$Q(\delta, L) = \sqrt{\frac{16\pi}{5}} \langle \delta, LM = L | \hat{Q}_{20} | \delta, LM = L \rangle$$

$$= \sqrt{\frac{16\pi}{5}} \begin{pmatrix} L & 0 & L \\ -L & 0 & L \end{pmatrix} \langle \delta, LM = L || \hat{Q}_2 || \delta, LM = L \rangle \, . \tag{6.32}$$

6.8   Studies of the Parameters of the GCM

This section explains some basic physical properties of the potential and its
parameters. Figures 6.2 and 6.3 show some elementary potential energy

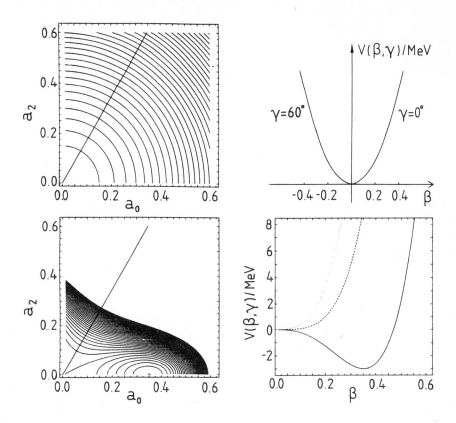

Fig. 6.2. Potential-energy surfaces for a spherical vibrator (*top*) and a prolate
rotor (*bottom*). On the left contours (distance: 0.5 MeV) are plotted, while the
right-hand graphs show cuts at constant γ. The cuts at 0 (30, 60) degrees are
plotted as *full (dashed, dotted) lines*.

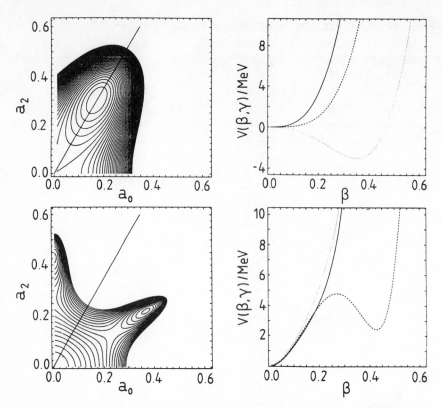

Fig. 6.3. Potential-energy surfaces for an oblate rotor (*top*) and a triaxial nucleus (*bottom*). Display is as Fig. 6.2.

surfaces (PESs), together with cuts at 0, 30, and 60 degrees. The upper part of Fig. 6.2 shows a harmonic-oscillator potential surface proportional to β^2. Thus C_2 is the only nonvanishing potential parameter in (6.9). This simplest PES characterizes a nucleus with a spherical ground state, performing harmonic quadrupole oscillations; an example is $^{120}_{52}$Te. The second row shows an axially symmetric deformed ground state on the prolate axis with the dominant excitation modes rotations and vibrations around that minimum. There are many examples for these *prolate rotors* in the mass region $A = 160 - 220$.

The upper part of Fig. 6.3 shows a similar case, but with the minimum on the oblate axis, while the last example corresponds to a nucleus with a spherical ground state, but also a secondary minimum at a triaxial deformation with all three principal axes of the nucleus unequal. In this case one would expect an additional set of states built on this excited triaxial shape and interposed on the vibrational spectrum.

Keeping in mind these elementary shapes, which already illustrate the flexibility of the approach, we now to discuss the effect of the several potential parameters in (6.9). C_2 generates a simple γ-independent potential, which is quadratic in β; one needs this potential type to generate harmonic oscillations. C_3 causes prolate–oblate differences in the PES: $C_3 > 0$ lowers the prolate half ($0 < \gamma < 30$) and raises the oblate half ($30 < \gamma < 60$). In conjunction with C_2 one can create a minimum on the prolate axis or on the oblate axis. C_4 and C_6 serve to create minima and determine the slope of the PES at higher β-values. C_5 has the same effect as C_3 except that it enlarges the PES at higher β-values. If $C_6 > 0$ the PES is raised increasingly with β for $\gamma \neq 30°$. Along $\gamma = 30°$ this contribution has a valley, and correspondingly a ridge may be generated at $\gamma = 30°$ for $C_6 < 0$. Together with C_2 and C_3 one can create asymmetric shapes that can be shifted to higher or lower γ-values with the help of C_3. The kinetic energy has two parameters: the mass parameter B_2 stretches the bands if it is lowered (and vice versa), while P_3 changes the relative position of the bands. In general, though, the effects of the anharmonic terms in the kinetic energy are less physically intuitive.

6.9   Fitting Experimental Data

The code allows an automatic fitting of a set of experimental data by adjusting the parameters of the Hamiltonian. The data may include energies, B(E2)-values, quadrupole moments, and energy differences, with the latter being useful for obtaining a better fit to certain crucial level spacings if desired. A least-squares-fit routine is needed for this purpose, for which we chose "ZXSSQ" from the IMSL library. The function minimized is called "FUNC". The states are characterized by their angular momentum L and sequence number N for fixed angular momentum starting at the lowest state. Thus, the yrast 2^+-state has $L = 2$ and $N = 1$, while the head of the β-band will usually have $L = 0$ and $N = 2$, as it is the second 0^+-state.

The experimental quantities are read into arrays with self-explanatory names. The level energies, for example, go into the array EXEN(i) (experimental energy), their angular momenta into LEXEN(i), and their sequence numbers into NEXEN(i). The code must, of course, know the total number of such data to be read, and this is provided in NUMEN, so that they are counted with $i = 1, \ldots$ NUMEN.

For the $B(E2)$-values the procedure is similar, except that each of these involves two states, so that a pair LEXBE2(i,1), NEXBE2(i,1) and LEXBE2(i,2), NEXBE2(i,2) is read together with the value itself, EXBE2(i). The number of $B(E2)$-values read is NUMBE2. The quadrupole values are read in complete analogy to the energies, using LEXQ(i), NEXQ(i), and EXQ(i), $i = \ldots$ NUMQ. In each case, the $B(E2)$-value is for the transition from the first state given to the second one.

In contrast, for the energy differences the sequence numbers in the whole list of levels $1 \ldots$ NUMEN are used to determine the states involved. The arrays are NEXDIF(i,1), NEXDIF(i,2), and EXDIF(i), $i = 1 \ldots$ NUMDIF.

The units used are: energies and their differences in MeV, $B(E2)$-values

Table 6.1. The experimental data of ^{186}Os used in the sample input file.

Level	Code numbering	Energy [MeV]	Quadrupole moment [eb]
0^+_{gs}	0^+_1	0.000	
2^+_{gs}	2^+_1	0.137	-1.55
4^+_{gs}	4^+_1	0.433	
6^+_{gs}	6^+_1	0.867	
2^+_γ	2^+_2	0.767	
3^+_γ	3^+_1	0.910	
4^+_γ	4^+_2	1.070	
5^+_γ	5^+_1	1.275	
6^+_γ	6^+_2	1.492	
4^+	4^+_3	1.352	

Transition	B(E2)-value $[e^2b^2]$
$0^+_{gs} \to 2^+_{gs}$	3.11
$2^+_{gs} \to 2^+_\gamma$	0.082
$2^+_{gs} \to 4^+_{gs}$	1.51

in e^2b^2, and quadrupole moments in eb. The actual fit done by the routine ZXSSQ minimizes the quantity

$$S = \sum_{i=0}^{NN} f_i^2,$$

where NN=NUMEN+NUMBE2+NUMQ+NUMDIF, and f_i is the deviation of the calculated quantity from the experimental one, suitably weighted. The subroutine FUNC passes these to ZXSSQ in the array F. Many alternative minimization schemes will use the sum S directly and not the individual deviations, so that in these cases (and usually also because of different argument conventions) FUNC should be modified accordingly.

For energies, quadrupole moments, and energy differences the weighting factors included in the code have been chosen from experience; they can easily be changed or even be read in independently for each piece of data. The factors at present are the same within each group of experimental data, and for the $B(E2)$-values a more complicated approach was chosen because of the larger discrepancies in magnitude found there. The natural logarithms of the $B(E2)$-values are used for the fit if $0.05 < B(E2) < 0.3$, and for values outside this range the absolute difference was retained to avoid excessive weighting of deviations in very small or very large $B(E2)$-values.

In general one should not expect miracles from such an automatic fitting and must scrutinize the results critically. The code searches for a minimum in an eight-dimensional parameter space, and there will usually be many subsidiary minima. The best way is to try different starting Hamiltonians and also to modify the weight in subroutine FUNC if the results are still not acceptable. The physicist's eye often finds certain features of the fit undesirable which the narrow-minded least-squares fit prefers; the only way to circumvent this is to decide what should be reproduced more accurately at the expense of other quantities, and to bias the fit accordingly. Sometimes it is even helpful to temporarily remove some experimental data from the fitting and only reintroduce them later when the fit has stabilized in the desired direction.

Finally, we give two useful tricks in adjusting the potential-energy surfaces to experimental data: multiplying all potential parameters by the same number does the same to the total PES and causes a common increase of the energy eigenvalues. This makes it possible to adjust an experimental energy value exactly. This rescaling factor is called RESCAL in the code and should be used after having finished a fit to get better agreement for the higher energies. In this case after each run of ENERG the energetically lowest state with highest angular momentum is automatically fitted; i.e., in general the $L = 6$-state of the ground-state band.

There is another trick that allows the exact adjustment of one transition probability: the canonical transformation

$$\alpha_{2m} \rightarrow \alpha'_{2m} = a\alpha_{2m}$$
$$\pi_{2m} \rightarrow \pi'_{2m} = \frac{1}{a}\pi_{2m}$$

leaves invariant the commutation relations of the collective coordinates and conjugate momenta, and thus does not change the spectrum of the Hamiltonian. The eigenfunctions also remain the same,

$$\psi(\alpha) \rightarrow \psi(\alpha').$$

However, there is a change of the matrix elements of the quadrupole operator (6.29); this operator corresponds to an observable and cannot be scaled arbitrarily. The matrix elements will thus be transformed according to

$$\langle\psi_1(\alpha)|\alpha_{2\mu}|\psi_2(\alpha)\rangle \rightarrow \langle\psi_1(\alpha')|\alpha_{2\mu}|\psi_2(\alpha')\rangle$$
$$= \frac{1}{a}\langle\psi_1(\alpha)|\alpha_{2\mu}|\psi_2(\alpha)\rangle$$
$$\equiv A_1 \qquad\qquad (6.33)$$

and

$$\langle\psi_1(\alpha)|\,[\alpha \times \alpha]_{[2\mu]}\,|\psi_2(\alpha)\rangle \rightarrow \langle\psi_1(\alpha')|\,[\alpha \times \alpha]_{[2\mu]}\,|\psi_2(\alpha')\rangle$$
$$= \frac{1}{a^2}\langle\psi_1(\alpha)|\,[\alpha \times \alpha]_{[2\mu]}\,|\psi_2(\alpha)\rangle$$
$$\equiv A_2 \qquad\qquad (6.34)$$

Thus, a certain expectation value $Q_{\exp}$ may be reproduced exactly by choosing the scaling factor a as

$$a = \frac{A_1}{2Q_{\exp}} \pm \sqrt{-\frac{A_2}{Q_{\exp}} + \frac{A_1^2}{4Q_{\exp}^2}}. \tag{6.35}$$

The positive value for a must be selected, because a negative a corresponds to a reflection at the 30-degree axis and interchanges oblate and prolate shapes. The code allows the use of this trick to obtain an exact fit of the $B(E2)$-value $0_1^+ \rightarrow 2_1^+$. This is controlled by the input variable IFFIT and its value is given by BE202. The option should be used with caution, though, because an exact fit of this one quantity may make the agreement of all other $B(E2)$-values much worse, and it may be preferable to demand an overall fit instead.

6.10   Finding the Optimal Basis Parameters

As a basis in this code we use the eigenfunctions of the five-dimensional harmonic oscillator, which are respectively parametrized in terms of the basis parameters C_2' and B_2'. For a finite basis dimension, the parameters have to be chosen to get satisfactory convergence of the calculated energies and $B(E2)$-values. To find the best set of basis parameters one has to diagonalize a given Hamiltonian and minimize the sum of (lowest) energy eigenvalues by varying the basis parameters (see e.g. Ref. [6.10]). Since this procedure is quite time-consuming, we use another scheme that takes much less time and turned out to be as effective: we minimize only the sum of the first NUM diagonal matrix elements of the Hamiltonian for spin $I = 0$ and take B_2' fixed at B_2. The integer variable NUM should equal the number of lowest $L = 0$ basis wave functions which contribute most to the first excited states. (Default: NUM = 10).

The minimum is found by increasing a do-loop variable S, defined as $S = (C_2' B_2' / \hbar^2)^{\frac{1}{4}}$, successively by 0.5. In case of failure to find reasonable basis parameters, the program is stopped and should be rerun with changed boundaries for S.

6.11   Radial and Angular Indices

Every wave function is defined uniquely by its quantum numbers. Instead of quantum numbers the code uses two indices for the definition of states: one for the radial part and one for the γ-dependent part. Both are calculated in the subroutine IAIR and are stored in the integer fields IA (Index Angular) and IR (Index Radial) respectively.

This makes the complete matrix elements a product of a radial and an angular part. Thus, for a given angular momentum L and phonon number ν, we have a one-to-one correspondence between the radial index and (ν, λ)

Table 6.2. The maximum number of allowed states (IMAX) for given angular momentum L and phonon number ν (see subroutine IAIR).

L	Phonon number ν					
	5	10	15	20	25	30
0	5	14	27	44	65	91
2	7	22	45	77	117	165
3	2	8	19	33	52	75
4	6	25	56	100	156	225
5	2	12	30	57	92	135
6	4	24	61	114	184	271
7	1	12	36	72	121	182
8	2	21	60	121	202	304
9	0	10	37	80	140	217
10	1	16	56	121	211	326

and the angular index and (μ, λ), respectively. The quantum numbers are subject to the selection rules (6.18), which are used in calculating the indices in IAIR. Finally, in Table 6.2, we show the number of possible basis states for a given phonon number and angular momentum.

6.12   Running the Code

For running the code additional standard subroutines are needed: a diagonalization subroutine used in ENERG (sufficient to use IFEB=1,2), a least-squares-fit routine used in FIT (needed for IFEB=4), and a linear-equation solver and a minimum routine used in POTMIN if one wants to calculate, e.g., the minimum or the stiffness in the β-direction of the PES from a given parameter set or vice versa.

In Table 6.3 we give an example input file, whose potential- and kinetic-energy data describe the nucleus ^{186}Os. This data set can be used as the standard input file, unit 5, in the code. All input variables are explained in the appendix. Table 6.1 shows the experimental values from ^{186}Os utilized for setting up the input data.

Besides the auxiliary routines the data files for the γ-matrix elements (unit numbers 20, 21) are needed as input, while the code produces an output file containing the diagonalization coefficients (unit number 30). Finally, for the fitting option, a file for reading the potential parameters is used (unit number 40).

6.13   List of Input Data

This section describes all the input data, which are read in free-field format from unit 5.

The data type follows the FORTRAN default: names starting with letters I through N denote integers and the others floating-point variables. For the arrangement of these data on the input lines, see the sample input file given in Table 6.3.

IFPARA: Switch determining the input method for the parameters in the Hamiltonian.

IFPARA=0: Read C2, C3, C4, C5, C6, D6, B2, P3 from unit 40. Ignore the values read from unit 5.

IFPARA=1: Read the same variables from standard input, unit 5.

IFPARA=2: Calculate C2–D6 from the physical parameters (next).

BETA0,GAMMA0,CB,CG,CBG,V0: Physical parameters, used only for IFPARA=2.

BETA0: β-value at the minimum of the PES.

GAMMA0: γ-value at the minimum of the PES.

CB: Stiffness in β-direction (i.e. $\frac{\partial^2 V(\beta,\gamma)}{\partial\beta^2}$) at the minimum.

CG: Stiffness in γ-direction (i.e. $\frac{\partial^2 V(\beta,\gamma)}{\partial\gamma^2}$) at the minimum.

CBG: Stiffness in (β,γ)-direction (i.e. $\frac{\partial^2 V(\beta,\gamma)}{\partial\beta\partial\gamma}$) at the minimum.

V0: Potential depth at the minimum.

C2, C3, C4, C5, C6 D6: Potential parameter given in (6.9) Units: MeV. Used only for IFPARA=1.

B2, P3: Kinetic-energy parameter given in (6.9). Units: B_2 in 10^{-42}MeVs^2 and P_3 in 10^{42}MeV/s^2. Used for IFPARA=1 or 2.

B2S, C2S: Basis parameters denoted in the text as B_2', C_2'.

IFBASE: Switch variable controlling whether the input basis parameters should be used as input (IFBASE=1) or whether the code should calculate a new basis parameter (IFBASE=0). (See subroutine SMIN).

NPH: Maximal phonon number. Determines the basis dimension (see Table 6.2).

IFEB: Switch variable controlling the type of calculation:

IFEB =1: Calculate energies, $B(E2)$ and quadrupole moments.

IFEB =2: Calculate only energies.

IFEB =4: Fitting experimental data.

LMINE, LMAXE: Minimum and maximum angular momentum for which the energies are calculated in ENERG.

NENEIV: Number of lowest-energy eigenvalues calculated in ENERG for every angular momentum.

RESCAL: Rescaling factor for the Hamiltonian providing an overall scaling factor for all energies. Recommended value: 1.

Table 6.3. Example input file for the nucleus ^{186}Os, containing the experimental data of Table 6.1.

0	IFPARA
0.,0.,0.,0.,0.,0.	BETA0,GAMMA0,CB,CG,CBG,V0
−564.76,733.01,13546.,−8535.1,−41635.,0.	C2,C3,C4,C5,C6,D6
112.48,−0.0531	B2,P3
90.,100.,0	B2S,C2S,IFBASE
30	NPH
1	IFEB
0,6	LMINE,LMAXE
3	NENEIV
1.0	RESCAL
3	NBQ
0,6	LMINB,LMAXB
186,76	NA,NZ
0.0	BE202
10,3,1,2	NUMEN,NUMBE2,NUMQ,NUMDIF
0	IFFIT
0	IFPLOT
0,1,0.0	LEXEN(I),NEXEN(I),EXEN(I)
2,1,0.137	...
2,2,0.767	...
3,1,0.910	...
4,1,0.433	...
4,2,0.070	...
4,3,0.352	...
5,1,0.275	...
6,1,0.867	...
6,2,0.492	...
0,1,2,1,3.11	(LEXBE2(I,K),NEXBE2(I,K),K=1,2),EXBE2(I)
2,1,2,2,0.082	...
2,1,4,1,1.51	...
2,1,−1.55	LEXQ(I),NEXQ(I),EXQ(I)
1,2,0.137	NEXDIF(I,1),NEXDIF(I,2),EXDIF(I)
2,3,0.630	...

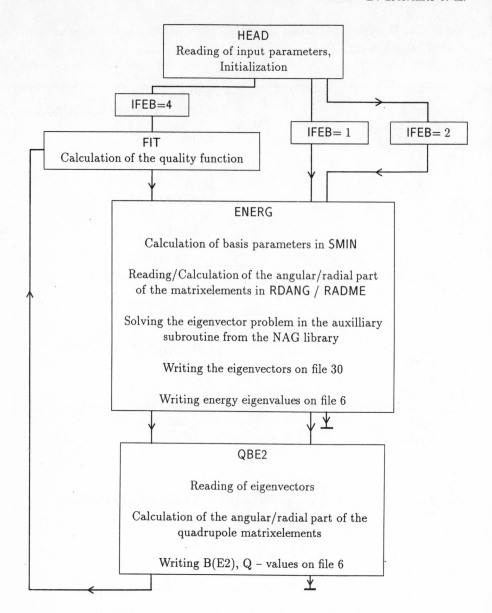

Fig. 6.4. Schematic flow chart of the whole GCM code.

NBQ: Number of states (for each angular momentum) for which all $B(E2)$s and quadrupole moments are calculated.

LMINB, LMAXB: Minimum and maximum angular momentum for which all $B(E2)$s and quadrupole moments are calculated. Together with NBQ, these allow a restriction of the expensive eigenvector calculations to the lower part of the spectrum.

NA, NZ: Mass and charge number of the nucleus considered.

BE202: The $B(E2; 0_1^+ \to 2_1^+)$ value which is to be adjusted exactly. If no adjustment is required, set BE202 $= 0$. (Units: e^2b^2.) LMINB must be equal to zero if BE202 is nonzero (In the code the variable BE202 is then used to store the square root of this number.)

NUMEN, NUMBE2, NUMQ, NUMDIF: Number of experimental energies, $B(E2)$-values, quadrupole moments, and energy differences to be read later in the input for fitting.

IFFIT: Switch variable deciding whether the energetically lowest state with highest angular momentum should be fitted exactly (IFFIT=1) or not (IFFIT=0). (Recommended value: 0.)

IFPLOT: Switch variable controlling whether, after each change in the Hamiltonian during fitting, cuts through the potential-energy surface should be plotted (IFPLOT=1) or not (IFPLOT=0). (Recommended value: 0.)

LEXEN(i), NEXEN(i), EXEN(i): Each of these lines contains an experimental energy value EXEN with the angular momentum LEXEN and the sequence number of the state in that angular-momentum group. The line "6,2,2.34", for example, says that the 2nd 6^+-state is at 2.34 MeV. The index i just counts the input lines for energies and is used similarly for the following sets of experimental data. The number of these lines must be NUMEN.

(LEXBE2(i,k), NEXBE2(i,k),k=1,2), EXBE2(i):
   Gives an experimental $B(E2)$-value (Units: e^2b^2 between the states defined by the angular momenta and indices LEXBE2 and NEXBE2. For example, the line "0,1,2,2,2.5" implies $B(E2, 0_1^+ \to 2_2^+) = 2.5\,e^2b^2$. The number of such lines must be equal to NUMBE2.

LEXQ(i), NEXQ(i), EXQ(i): Input line for an experimental quadrupole moment (Unit: eb) of the NEXQth state with angular momentum LEXQ. Thus "2,2,0.38" means $Q(2_2^+) = 0.38eb$. There must be NUMQ such lines.

NEXDIF(i,1), NEXDIF(i,2), EXDIF(i): Line for an experimental energy difference (in MeV) between the states numbered NEXDIF(i,1) and NEXDIF(i,2). These numbers refer to the number of the states in the *total* list of input energies, not for the angular momenta separately. For example, the line "2,3,0.630" in the sample input deck expresses the fact that $E(2_2^+) - E(2_1^+) = 0.630\,\text{MeV}$.

6.14   Technical Note

The program uses a few subroutines for standard mathematical calculations, most of them from the IMSL library. The following discussion should be sufficient to enable the user to substitute his own routines. More information can be found in GCM.FOR itself and in the IMSL library user's guide (Edition 9.2, (1984)).

GIVENS

Purpose:		Compute eigenvalues and eigenvectors of a real symmetric matrix. (Can be replaced by EIGRS from IMSL)
Usage:		CALL GIVENS(H,W,DIAGCO,IMAX,NENEIV,IDIAG)
Arguments:		
H	-	Matrixelements in band symmetric storage mode (input).
W	-	Energy eigenvalues in ascending order (output).
DIAGCO	-	Matrix storing the diagonalization coefficients of the n-th eigenvalue in the n-th row (output).
IMAX	-	Dimension of input matrix H (input).
NENEIV	-	Number of lowest energy eigenvalues to be calculated.
IDIAG	-	Maximum row dimension of DIAGCO (input).

ZXMWD (IMSL)

Purpose:		Global minimum of a function with N variables with constraints.
Usage:		CALL ZXMWD(FUNCT,N,NSIG,A,B,NSRCH,Y,F, WORK,IWORK,IER)
Arguments:		
FUNCT	-	Function subroutine which calculates F for given $X(1), \ldots, X(N)$.
N	-	Number of unknown parameters (input).
NSIG	-	Number of accuracy digits requiered for the parameter estimtates (input).
A,B	-	Constraint vectors of length N (input).
NSRCH	-	Number of starting points to be generated (input).
Y	-	Vector containg the final parameter estimates.
F	-	Function values at the final parameter estimates (output).
IER	-	Error parameter (output).

ZXSSQ (IMSL)

Purpose:		Minimum of the sum of squares of M functions in N variables.
Usage:		CALL ZXSSQ(FUNC,M,N,NSIG,EPS,DELTA, MAXFN,IOPT,PARM,X,SSQ,F,XJAC, IXJAC,XJTJ,WORK,INFER,IER)

Arguments:

FUNC	-	Subroutine calculating the residual vector $F(1), \ldots, F(M)$ for the given parameter values $X(1), \ldots, X(N)$.
M	-	Number of residuals, i.e. experimental data to be fitted (input).
N	-	Number of parameters (input).
NSIG,EPS,DELTA	-	Parameters for three convergence criteria (input).
MAXFN	-	Maximum number of function evaluations (input).
IOPT	-	Input options parameter to switch the minimization algorithm.
PARM	-	Input parameters only necessary for a certain choice of IOPT.
X	-	Vector of parameters containing the initial guess on input and the final parameter estimates of ZXSSQ on output.
SSQ	-	Residual sum of squares $F(1) \cdot F(1)+, \ldots, +F(M) \cdot F(M)$ (output).
F	-	Vector of residuals (output).
XJAC	-	Jacobian of output vector X (output).
IXJAC	-	Input row dimension of XJAC.
XJTJ	-	Vector containing $XJAC^T * XJAC$ on output.
WORK	-	Work vector.
INFER	-	Indicates if convergence was obtained (output).
IER	-	Error parameter.

LEQT2F (IMSL)

Purpose:	Linear equation solution in full storage mode with high accuracy.

Usage: CALL
                 LEQT2F(A,M,N,IA,B,IDGT,WKAREA,IER)
Arguments:
A   -   Input coefficient matrix of $Ax = B$ (full storage
                 mode).
M   -   Number of right hand sides (input).
N   -   Order of A (input).
IA   -   Row dimension of A (input).
B   -   Right hand side of $Ax = B$ (input).
IDGT   -   On input: Number of required accuracy digits.
                 On output: Approximate Number of obtained ac-
                 curacy digits.
WKAREA   -   Work vector.
IER   -   Error parameter (output).

References

[6.1] J.M. Eisenberg and W. Greiner, *Nuclear Theory Vol 1, Nuclear Models* 3rd
      edition (North-Holland, Amsterdam, 1987)
[6.2] P.O. Hess, Diploma Thesis, J.W. Goethe-Universität, Frankfurt/Main, Ger-
      many, 1978
[6.3] P.O. Hess, PhD Thesis, J.W. Goethe-Universität, Frankfurt/Main, Germany
      1980.
[6.4] P.O. Hess, M. Seiwert, J. Maruhn, and W. Greiner Z. Physik **A296** (1980)
      147
[6.5] E. Chacon and M. Moshinsky, J. Math. Phys., **17** No.5 (1976) 668
[6.6] E. Chacon and M. Moshinsky J. Math. Phys., **18** No.5 (1977) 870
[6.7] A. Bohr Kgl. Danske Videnskab Selskab Mat.-Fys. Medd. **26**, 14 (1952)
[6.8] G. Gneuss and W. Greiner, Nucl. Phys. **A171** (1971) 449
[6.9] D.A. Varshalovich and A.N. Moskalev, V.K. Khersonskii *Quantum Theory of
      Angular Momentum* (World Scientific, Singapore 1988)
[6.10] P.O. Löwdin, Phys. Rev. **139** No.2A (1965)

7. The Relativistic Impulse Approximation

C. J. Horowitz, D. P. Murdock, and Brian D. Serot

7.1 Background

The scattering of medium-energy nucleons from nuclei can provide information about both nuclear structure and the nucleon–nucleon (NN) interaction. Since the NN interaction has complex spin, isospin, momentum, and density dependence, nucleon–nucleus scattering exhibits a variety of phenomena. As a starting point for describing these phenomena, we use the impulse approximation, which assumes that the interaction between the projectile and target nucleons has essentially the same form as the interaction between two nucleons in free space. This approximation also implies that the scattering is dominated by single-collision processes, so it is most appropriate for scattering at medium energies and small scattering angles.

Nuclear structure and the NN interaction have been studied for many years within the context of the nonrelativistic Schrödinger equation. There is a growing body of evidence, however, that the NN interaction contains large Lorentz scalar and four-vector components. To maintain the distinction between these components, it is necessary to use the Dirac equation. Thus we will describe the motion of the projectile in the field of the nucleus with the Dirac equation, and to be consistent, we also compute the target structure in terms of Dirac nucleons.

The basic ingredients contained in the codes that follow are the ground-state nuclear wave function, computed in the Dirac–Hartree approximation, and the Dirac optical potential for the projectile, computed in the so-called relativistic impulse approximation (RIA). The target nucleons interact with each other through static mean fields of Lorentz scalar and vector character; these fields are produced by the scalar and number densities of the nucleons, so that the ground-state calculation is self-consistent. For simplicity, we consider only spherically symmetric ("doubly magic") nuclei with well-closed shells.

Once the nuclear ground state is determined, the densities are folded with a relativistic parametrization of the free-space NN interaction to produce optical potentials that enter in the Dirac equation for the projectile. This equation is then solved to compute the observables for proton scattering from a spin-zero nucleus: the differential cross section ($d\sigma/d\Omega$), the analyzing power (A_y), and the spin-rotation function (Q). Note that while the ground-state equations contain several parameters that are adjusted to reproduce observed bulk nuclear properties, the parametrization of the NN interaction is just a convenient way of including the empirically measured scattering amplitudes. For projectile kinetic energies below 500 MeV in the laboratory, however, the free NN amplitudes must be modified to allow for Pauli blocking and exchange corrections, as we describe below, in order to produce meaningful results.

General background for the relativistic description of nuclear structure can be found in Refs. [7.1,2], and references cited therein. The Dirac–Hartree approximation for spherical nuclei is described in Refs. [7.3–9], and extensions for nonspherical nuclei are discussed in Refs. [7.13–16] to [7.12]. The original RIA was formulated in refs. [7.13] to [7.16], after being motivated by numerous earlier studies of proton–nucleus scattering using phenomenological Dirac potentials [7.17]. Various modifications to the original RIA, which are necessary to apply this approach at lower projectile energies, have subsequently been developed [7.18–20]. The analysis described here follows closely the work in Refs. [7.7,18,19].

7.2   The Dirac–Hartree Approximation

The Dirac–Hartree equations for a finite nucleus can be derived from an interacting relativistic field theory of mesons and baryons by approximating the meson field operators by classical fields. In what follows, we will exhibit only the contributions from the neutral scalar (ϕ) and vector (V^μ) meson fields, as in the Walecka model [7.1], but the code also contains contributions from a neutral (isovector) ρ meson and the coulomb potential [7.7].

If we restrict consideration to static, spherically symmetric nuclei, the meson fields depend only on the radius, and only the V^0 component of the vector field contributes. Thus the Dirac equation for the baryon field (ψ) is

$$\{i\gamma_\mu\partial^\mu - g_v\gamma^0 V^0(r) - [M - g_s\phi(r)]\}\psi(x) = 0\,. \tag{7.1}$$

Here the γ^μ matrices are defined as by Bjorken and Drell [7.21], and appropriate values for the scalar and vector couplings g_s and g_v will be given below. Although the baryon field is still an operator, the meson fields are classical; hence (7.1) is linear, and we may seek normal-mode solutions of the form $\psi(x) = \psi(\boldsymbol{x})e^{-iEt}$. This leads to

$$h\psi(\boldsymbol{x}) \;=\; E\psi(\boldsymbol{x})\,, \tag{7.2}$$
$$h \;\equiv\; -i\boldsymbol{\alpha}\cdot\boldsymbol{\nabla} + g_v V^0(r) + \beta[M - g_s\phi(r)]\,, \tag{7.3}$$

which defines the single-particle Dirac Hamiltonian h. Equation (7.2) has both positive- and negative-energy solutions $\mathcal{U}(\boldsymbol{x})$ and $\mathcal{V}(\boldsymbol{x})$, and thus the field operator can be expanded as

$$\hat{\psi}(\boldsymbol{x}) = \sum_\alpha \left[A_\alpha \mathcal{U}_\alpha(\boldsymbol{x}) + B_\alpha^\dagger \mathcal{V}_\alpha(\boldsymbol{x}) \right] \tag{7.4}$$

in the Schrödinger picture. Here $A_\alpha^\dagger$ and $B_\alpha^\dagger$ are baryon and antibaryon creation operators that satisfy the standard anticommutation relations. The label α specifies the full set of single-particle quantum numbers, and since the system is assumed spherically symmetric and parity conserving, α contains the usual angular-momentum and parity quantum numbers, as described in Refs. [7.21] and [7.2].

Using well-known properties of the relativistic angular-momentum operators, it is easy to show that the angular and spin solutions are spin spherical harmonics [7.22]:

$$\Phi_{\kappa m} = \sum_{m_\ell m_s} \langle \ell m_\ell \tfrac{1}{2} m_s | \ell \tfrac{1}{2} j m \rangle Y_{\ell,m_\ell}(\Omega) \chi_{m_s}, \tag{7.5}$$

$$j = |\kappa| - \tfrac{1}{2}, \qquad \ell = \begin{cases} \kappa, & \kappa > 0, \\ -(\kappa+1), & \kappa < 0, \end{cases} \tag{7.6}$$

where Y_{ℓ,m_ℓ} is a spherical harmonic, χ_{m_s} is a two-component Pauli spinor, and κ is a nonzero integer. Notice that κ uniquely defines j and ℓ, and it follows that the four-component Dirac wave functions can be divided into upper and lower two-component pieces with opposite values of κ. Thus, the positive-energy spinors can be written as

$$\mathcal{U}_\alpha(\boldsymbol{x}) \equiv \mathcal{U}_{n\kappa m t}(\boldsymbol{x}) = \begin{pmatrix} i[G_{n\kappa t}(r)/r]\,\Phi_{\kappa m} \\ -[F_{n\kappa t}(r)/r]\,\Phi_{-\kappa m} \end{pmatrix} \zeta_t, \tag{7.7}$$

where n is the principal quantum number and ζ_t is a two-component isospinor labeled by the isospin projection t. The phase choice in (7.7) produces real bound-state wave functions F and G for real potentials ϕ and V^0, and the normalization is given by

$$\int_0^\infty dr \left(|G_\alpha(r)|^2 + |F_\alpha(r)|^2 \right) = 1. \tag{7.8}$$

With the general form for the spinors in (7.7), we can evaluate the nuclear densities, which serve as source terms in the meson field equations. Assume that the nuclear ground state consists of filled shells up to some value of n and κ, which may be different for protons and neutrons; this is appropriate for doubly magic nuclei. In addition, assume that all bilinear products of baryon operators are normal ordered, which removes contributions from the negative-energy spinors $\mathcal{V}_\alpha(\boldsymbol{x})$. This amounts to neglecting contributions from the filled Dirac sea of baryons. These contributions can be included if desired [7.9,23–25], but are beyond the scope of the present development.

With these assumptions, the local baryon (ρ_B) and scalar (ρ_s) densities become

$$\left. \begin{matrix} \rho_B(\boldsymbol{x}) \\ \rho_s(\boldsymbol{x}) \end{matrix} \right\} = \sum_\alpha^{\text{occ}} \overline{\mathcal{U}}_\alpha(\boldsymbol{x}) \begin{pmatrix} \gamma^0 \\ 1 \end{pmatrix} \mathcal{U}_\alpha(\boldsymbol{x})$$

$$= \sum_\alpha^{\text{occ}} \left(\frac{2j_a + 1}{4\pi r^2} \right) \left(|G_a(r)|^2 \pm |F_a(r)|^2 \right), \tag{7.9}$$

where we have used

$$\sum_{m=-j}^{j} \Phi^{\dagger}_{\kappa m} \Phi_{\kappa' m} = \left(\frac{2j+1}{4\pi}\right)\delta_{\kappa\kappa'}, \qquad \kappa' = \pm\kappa, \tag{7.10}$$

which holds for filled shells, and the remaining quantum numbers are denoted by $\{\alpha\} = \{a; m\} \equiv \{n, \kappa, t; m\}$. Note that since the shells are filled, the sources are spherically symmetric. The sources produce the meson fields, which satisfy static Klein–Gordon equations:

$$\frac{d^2}{dr^2}\phi(r) + \frac{2}{r}\frac{d}{dr}\phi(r) - m_s^2\phi(r) = -g_s\rho_s(r), \tag{7.11}$$

$$\frac{d^2}{dr^2}V^0(r) + \frac{2}{r}\frac{d}{dr}V^0(r) - m_v^2V^0(r) = -g_v\rho_B(r). \tag{7.12}$$

Here m_v and m_s are the vector and scalar meson masses. For the Coulomb potential, one uses the contribution to ρ_B arising from protons only, while for the ρ meson, one uses half the difference between the proton and neutron densities [7.7].

The equations for the baryon wave functions follow upon substituting (7.7) into (7.2), which produces

$$\frac{d}{dr}G_a(r) + \frac{\kappa}{r}G_a(r) - \left[E_a - g_vV^0(r) + M - g_s\phi(r)\right]F_a(r) = 0, \tag{7.13}$$

$$\frac{d}{dr}F_a(r) - \frac{\kappa}{r}F_a(r) + \left[E_a - g_vV^0(r) - M + g_s\phi(r)\right]G_a(r) = 0. \tag{7.14}$$

Thus the spherical nuclear ground state is described by coupled, one-dimensional differential equations that may be solved by an iterative procedure, as we describe below. Once the solution has been found, the total energy of the system is given by

$$E = \sum_{a}^{occ} E_a(2j_a + 1) - \frac{1}{2}\int d^3x\left[-g_s\phi(r)\rho_s(r) + g_vV^0(r)\rho_B(r)\right]. \tag{7.15}$$

The solutions of the preceding equations depend on the parameters g_s, g_v, m_s, and g_ρ (when the ρ meson is included). We take the experimental values $M = 939$ MeV, $m_v = m_\omega = 783$ MeV, $m_\rho = 770$ MeV, and $e^2/4\pi = \alpha = 1/137.036$ (which determines the Coulomb potential) as fixed. The free parameters can be chosen by requiring that when the Dirac–Hartree equations are solved in the limit of infinite nuclear matter, the empirical equilibrium density ($\rho_B^0 = 0.1484$ fm^{-3}), binding energy (15.75 MeV), and symmetry energy (35 MeV) are reproduced. Moreover, we fit the empirical rms charge radius of ^{40}Ca ($r_{rms} = 3.482$ fm), which is determined primarily by m_s. This procedure results in the following parameter values:

$$g_s^2 = 109.6, \quad g_v^2 = 190.4, \quad g_\rho^2 = 65.23, \quad m_s = 520 \text{ MeV}. \tag{7.16}$$

These values are used as input in the Dirac–Hartree code TIMORA, and can be changed if desired.

7.3   The Nucleon–Nucleon Scattering Amplitude

The original RIA involves two basic procedures [7.14–16]. First, the experimental NN scattering amplitude is represented by a particular set of five Lorentz covariant functions [7.26], which multiply the so-called Fermi-invariant Dirac matrices. The Lorentz-covariant functions are then folded with the target densities to produce a first-order optical potential for use in the Dirac equation for the projectile.

The use of the Fermi invariants involves implicit dynamical assumptions about the NN interaction, and this choice is not unique. Moreover, the standard representation of the amplitude has somewhat awkward behavior under the interchange of the two (identical) nucleons, which produces difficulties in treating the exchange contributions to nucleon–nucleus scattering. These difficulties grow more serious at low energies and large scattering angles. Finally, the original RIA assumes that the NN interaction between projectile and target is unmodified by the surrounding nucleons. While this assumption is valid at high energies, there are important modifications from Pauli blocking at lower energies, which we will include here in an approximate fashion. There may also be important binding-energy corrections from the nuclear medium, but these are not yet well understood and are omitted here.

The constraints of Lorentz covariance, parity conservation, isospin invariance, and that free nucleons are on their mass shell imply that the invariant NN scattering operator $\mathcal{F}$ can be written in terms of five complex functions for pp scattering and five for pn scattering. In the original RIA, $\mathcal{F}$ was taken as

$$\mathcal{F} = \mathcal{F}^S + \mathcal{F}^V \gamma^\mu_{(0)} \gamma_{(1)\mu} + \mathcal{F}^{PS} \gamma^5_{(0)} \gamma^5_{(1)} + \mathcal{F}^T \sigma^{\mu\nu}_{(0)} \sigma_{(1)\mu\nu} + \mathcal{F}^A \gamma^5_{(0)} \gamma^\mu_{(0)} \gamma^5_{(1)} \gamma_{(1)\mu} \,, \quad (7.17)$$

where the subscripts (0) and (1) refer to the incident and struck nucleons, respectively. Each amplitude $\mathcal{F}^L$ is a complex function of the Lorentz invariants t (four-momentum transfer squared) and s (total four-momentum squared), or equivalently, of the momentum transfer q and laboratory energy E. Thus we may rewrite (7.17) in the convenient form

$$\mathcal{F}(q, E) = \sum_L \mathcal{F}^L(q, E) \, \lambda^L_{(0)} \cdot \lambda^L_{(1)} \,, \quad (7.18)$$

where the $\lambda^L_{(i)}$ stand for the Dirac operators for the incident and struck nucleons, and the dot product implies that all Lorentz indices are contracted, as in (7.17).

The NN scattering amplitude is obtained by taking matrix elements of $\mathcal{F}$ between initial and final two-nucleon states described by Dirac spinors. In practice, the functions $\mathcal{F}^L$ are determined by equating the resulting amplitude (in the center-of-mass frame) to the empirical amplitude, which is conventionally expressed in terms of the nonrelativistic Wolfenstein amplitudes $A, \ldots, E$. Since there are five complex invariant amplitudes and five complex Wolfenstein amplitudes, the relationship is determined by a 5×5 nonsingular matrix, whose inversion is straightforward [7.26].

We emphasize, however, that $\mathcal{F}$ is an *operator* in the two-particle Dirac space; it has 256 components when taken between all combinations of nucleon and antinucleon spinors at a given q and E. Although isospin and parity invariance reduce this to 44 components [7.20], the procedure described above determines only five (complex) functions, which is not sufficient to specify $\mathcal{F}$ uniquely. In other words, there are an infinite number of operators $\mathcal{F}$ with the same five on-shell, positive-energy matrix elements, but different negative-energy matrix elements. This is relevant because the *full* $\mathcal{F}$, and not just its on-shell, positive-energy matrix elements, is used to construct the nucleon–nucleus optical potential. Thus there are nontrivial assumptions involved in using (7.18) to compute the optical potential. Additional physical arguments must be used to choose the most meaningful λ^L, since these determine the behavior of the NN amplitude when the spinors change in the nuclear medium. In particular, although 7.17 may be acceptable at high energies, it is now known that the pseudoscalar piece $\mathcal{F}^{PS}$ should be replaced by the pseudovector invariant

$$\mathcal{F}^{PS}\gamma_{(0)}^5 \gamma_{(1)}^5 \rightarrow -\mathcal{F}^{PV} \frac{\not{q}\gamma_{(0)}^5}{2M} \frac{\not{q}\gamma_{(1)}^5}{2M} \tag{7.19}$$

to give meaningful results at lower energies, as we describe below.

In addition to the problem of nonuniqueness of negative-energy matrix elements, the original RIA procedure does not separate the direct and exchange contributions to the amplitude. These have different characteristic dependence on the momentum transfer and energy, which becomes important at low energies. To overcome this problem, a model was constructed in Ref. [7.18] that describes the NN amplitude using direct and exchange Feynman diagrams containing a set of "mesons". The mesons have different spins and parities (scalar, vector, tensor, pseudoscalar, and axial vector) and isospin 0 or 1. The meson–nucleon couplings are complex, with a real part g_i^2 and an imaginary one $\bar{g}_i^2$. The small imaginary couplings are a phenomenological means of obtaining the imaginary part of the NN amplitude. This model decomposes the original RIA amplitude into

$$\langle k_0' k_1' | \mathcal{F} | k_0 k_1 \rangle = \langle k_0' k_1' | t(E) | k_0 k_1 \rangle + (-1)^T \langle k_1' k_0' | t(E) | k_0 k_1 \rangle, \tag{7.20}$$

where $t(E)$ is the lowest-order "meson" exchange diagram evaluated from the Feynman rules, and T is the total isospin of the two-nucleon state.

The calculation of the one-meson-exchange Feynman diagrams is straightforward, and we assume that the vertices carry form factors, so that they take the form

$$g_i \left(\frac{\Lambda_i^2}{q^2 + \Lambda_i^2} \right) \lambda^{L(i)}(\tau)^{I_i}, \tag{7.21}$$

where $L(i)$ denotes the spin and parity of the ith meson and $I_i = (0,1)$ is the meson's isospin. We neglect the energy transfer q^0 carried by the meson and allow for different meson masses and cutoff parameters in the real and imaginary parts of the amplitude [see (7.28)]. Then,) up to overall kinematic factors, the contribution of meson i to the NN scattering amplitude is

$$\overline{U}_{0'}\overline{U}_{1'}\mathcal{F}_i U_0 U_1 \propto \frac{g_i^2}{q^2 + m_i^2}\left(\frac{\Lambda_i^2}{\Lambda_i^2 + q^2}\right)^2 \{\boldsymbol{\tau}_0 \cdot \boldsymbol{\tau}_1\}^{I_i} \overline{U}_{0'}\lambda^{L(i)}U_0 \cdot \overline{U}_{1'}\lambda^{L(i)}U_1$$

$$+ (-1)^T \frac{g_i^2}{Q^2 + m_i^2}\left(\frac{\Lambda_i^2}{\Lambda_i^2 + Q^2}\right)^2 \{\boldsymbol{\tau}_0 \cdot \boldsymbol{\tau}_1\}^{I_i} \overline{U}_{1'}\lambda^{L(i)}U_0 \cdot \overline{U}_{0'}\lambda^{L(i)}U_1 ,$$

$$(7.22)$$

where the direct and exchange momentum transfers are $\boldsymbol{q} = \boldsymbol{k}_0' - \boldsymbol{k}_0$ and $\boldsymbol{Q} = \boldsymbol{k}_1' - \boldsymbol{k}_0$.

The first term in (7.22) is already of the form in (7.18), so that we can immediately identify the contributions to $\mathcal{F}^L$. The second term, however, is not of this form, but it can be rewritten using a Fierz transformation, with the result

$$\overline{U}_0\overline{U}_{1'}\mathcal{F}_i U_0 U_1 \propto \frac{g_i^2}{q^2 + m_i^2}\left(\frac{\Lambda_i^2}{\Lambda_i^2 + q^2}\right)^2 \{\boldsymbol{\tau}_0 \cdot \boldsymbol{\tau}_1\}^{I_i} \overline{U}_{0'}\lambda^{L(i)}U_0 \cdot \overline{U}_{1'}\lambda^{L(i)}U_1$$

$$+ (-1)^T \frac{g_i^2}{Q^2 + m_i^2}\left(\frac{\Lambda_i^2}{\Lambda_i^2 + Q^2}\right)^2 \{\boldsymbol{\tau}_0 \cdot \boldsymbol{\tau}_1\}^{I_i} \sum_{L'} B_{L(i),L'}\overline{U}_{0'}\lambda^{L'}U_0 \cdot \overline{U}_{1'}\lambda^{L'}U_1 ,$$

$$(7.23)$$

where the transformation matrix is given by

$$B_{L,L'} \equiv \frac{1}{8}\begin{pmatrix} 2 & 2 & 1 & -2 & 2 \\ 8 & -4 & 0 & -4 & -8 \\ 24 & 0 & -4 & 0 & 24 \\ -8 & -4 & 0 & -4 & 8 \\ 2 & -2 & 1 & 2 & 2 \end{pmatrix} . \qquad (7.24)$$

(The rows and columns are labeled in the order S, V, T, A, PS.)

It is now simply a matter of algebra to compute the correct overall kinematic factors that determine the contributions to the Lorentz invariants $\mathcal{F}^L$. These may be written as

$$\mathcal{F}^L(q, E_c) = i\frac{M^2}{2E_c k_c}\left[F_D^L(q) + F_X^L(Q)\right], \qquad (7.25)$$

$$F_D^L(q) \equiv \sum_i \delta_{L,L(i)}\{\boldsymbol{\tau}_0 \cdot \boldsymbol{\tau}_1\}^{I_i} f^i(q), \qquad (7.26)$$

$$F_X^L(Q) \equiv (-1)^T \sum_i B_{L(i),L}\{\boldsymbol{\tau}_0 \cdot \boldsymbol{\tau}_1\}^{I_i} f^i(Q), \qquad (7.27)$$

$$f^i(q) \equiv \frac{g_i^2}{q^2 + m_i^2}\left(\frac{\Lambda_i^2}{\Lambda_i^2 + q^2}\right)^2 - i\frac{\bar{g}_i^2}{q^2 + \bar{m}_i^2}\left(\frac{\bar{\Lambda}_i^2}{\bar{\Lambda}_i^2 + q^2}\right)^2 . \qquad (7.28)$$

Here E_c is the total energy and k_c is the relative momentum in the nucleon–nucleon cm system. Note that the f^i depend only on the magnitude of the three-momentum transfer, which we write simply as q or Q.

For pp scattering, one uses only the $T = 1$ parts of the amplitude, while for pn scattering, the average of the $T = 1$ and $T = 0$ parts are required. These expressions were used in Ref. [7.18] to fit the NN scattering amplitudes at several different laboratory scattering energies. The full set of parameters can be found in Refs. [7.18,19]. Note that using a derivative coupling for the pseudoscalar meson, as in (7.19), does not change the values obtained from the fit, because on-shell spinor matrix elements are identical for pseudoscalar and pseudovector operators.

7.4   The Nucleon–Nucleus Optical Potential

In the original RIA, the Dirac optical potential $U_{\mathrm{opt}}(q, E)$ can be written as

$$U_{\mathrm{opt}}(q, E) = -\frac{4\pi i p}{M} \langle \Psi | \sum_{n=1}^{A} e^{iq \cdot x(n)} \mathcal{F}(q, E; n) | \Psi \rangle , \qquad (7.29)$$

where $\mathcal{F}$ is the scattering operator of (7.17), p is the magnitude of the three-momentum of the projectile in the nucleon–*nucleus* cm frame (where the scattering observables will be calculated), and $|\Psi\rangle$ is the A-particle nuclear ground state. $\mathcal{F}$ is a function of the momentum transfer q and collision energy E, which we take to be the cm energy for a stationary target nucleon and incident proton at the laboratory energy; this amounts to neglecting nuclear recoil.

The operator $\mathcal{F}(q, E; n)$ describes the scattering of the projectile from target nucleon n without a separation into direct and exchange terms. With the model described above, however, these terms can be included explicitly, to provide an improved expression for the optical potential. If we denote the incident projectile wave function as $\mathcal{U}_0(x)$ and assume that the nuclear ground state is described by a Dirac–Hartree wave function, the action of the optical potential on the incident wave, projected into coordinate space, can be written as [7.27]

$$\langle x | U_{\mathrm{opt}} | \mathcal{U}_0 \rangle = -\frac{4\pi i p}{M} \sum_{\alpha}^{\mathrm{occ}} \int \mathrm{d}^3 y' \, \mathrm{d}^3 y \, \mathrm{d}^3 x' \, \overline{\mathcal{U}}_\alpha(y')$$

$$\times \left\{ \langle xy' | t(E) | x'y \rangle + (-1)^T \langle y'x | t(E) | x'y \rangle \right\} \mathcal{U}_0(x') \mathcal{U}_\alpha(y) . \quad (7.30)$$

The antisymmetrized matrix element of $t(E)$ in coordinate space is the Fourier transform of the momentum-space matrix element, for which the model discussed in Sect. 7.3 gives explicit expressions. The Fierz transformation can be used to express $t(E)$ in the form of (7.18). (Subtleties associated with the use of the *PV* vertex will be considered below.) After taking the required Fourier transforms, one finds [7.27]

$$\langle \boldsymbol{x}|U_{\rm opt}|\mathcal{U}_0\rangle = -\frac{4\pi i p}{M}\sum_L \left[\int {\rm d}^3 x'\rho^L(\boldsymbol{x}')t_D^L(|\boldsymbol{x}'-\boldsymbol{x}|;E)\right]\lambda^L\mathcal{U}_0(\boldsymbol{x})$$

$$-\frac{4\pi i p}{M}\sum_L \int {\rm d}^3 x'\rho^L(\boldsymbol{x}',\boldsymbol{x})t_X^L(|\boldsymbol{x}'-\boldsymbol{x}|;E)\lambda^L\mathcal{U}_0(\boldsymbol{x}')\,(7.31)$$

where

$$t_D^L(|\boldsymbol{x}|;E) \equiv \int \frac{{\rm d}^3 q}{(2\pi)^3}\,t_D^L(q,E){\rm e}^{-iq\cdot x}\,, \qquad (7.32)$$

with $t_D^L(q,E) \equiv (iM^2/2E_ck_c)F_D^L(q)$ and similarly for the exchange pieces $t_X^L(Q,E)$. The nuclear densities are defined as in (7.9) :

$$\rho^L(\boldsymbol{x}',\boldsymbol{x}) \equiv \sum_\alpha^{\rm occ'} \overline{\mathcal{U}}_\alpha(\boldsymbol{x}')\lambda^L\mathcal{U}_\alpha(\boldsymbol{x})\,, \qquad \rho^L(\boldsymbol{x}) \equiv \rho^L(\boldsymbol{x},\boldsymbol{x})\,. \qquad (7.33)$$

Here the prime on the occupied states means that one sums over target protons when the density is to be used with *pp* amplitudes and over target neutrons when it is to be used with *pn* amplitudes.

The first term in (7.31) contains a multiplicative factor that defines the direct optical potential

$$U_D^L(r;E) \equiv -\frac{4\pi i p}{M}\int {\rm d}^3 x'\rho^L(\boldsymbol{x}')t_D^L(|\boldsymbol{x}'-\boldsymbol{x}|;E)\,. \qquad (7.34)$$

The second (exchange) term is nonlocal, but is treated in the local-density approximation of Brieva and Rook [7.28], which uses plane-wave states for the incident and bound nucleons. This allows us to replace the exchange integral with a local potential given by

$$U_X^L(r;E) \equiv -\frac{4\pi i p}{M}\int {\rm d}^3 x'\rho^L(\boldsymbol{x}',\boldsymbol{x})t_X^L(|\boldsymbol{x}'-\boldsymbol{x}|;E)j_0(p|\boldsymbol{x}'-\boldsymbol{x}|)\,, \qquad (7.35)$$

where j_0 is a spherical Bessel function. The off-diagonal one-body density is approximated by the local-density result

$$\rho^L(\boldsymbol{x}',\boldsymbol{x}) \approx \rho^L\left(\tfrac{1}{2}(\boldsymbol{x}+\boldsymbol{x}')\right)\left(\frac{3}{sk_{\rm F}}\right)j_1(sk_{\rm F})\,, \qquad (7.36)$$

where $s \equiv |\boldsymbol{x}'-\boldsymbol{x}|$ and $k_{\rm F}$ is related to the nuclear baryon density by $\rho_{\rm B}\left(\tfrac{1}{2}(\boldsymbol{x}+\boldsymbol{x}')\right) = 2k_{\rm F}^3/3\pi^2$.

With these simplifications, the Dirac optical potential is local, and only diagonal nuclear densities are needed. For a spin-zero nucleus, the only nonzero densities are the baryon and scalar densities of (7.9), and a tensor term computed by inserting σ^{0i} between the spinors in (7.9). Thus, the optical potential takes the form

$$U_{\text{opt}} = U^S + \gamma^0 U^V - 2i\boldsymbol{\alpha}\cdot\hat{r}U^T, \tag{7.37}$$

where

$$U^L \equiv U^L(r; E) = U_D^L(r; E) + U_X^L(r; E). \tag{7.38}$$

The tensor contribution is found to be small and will be neglected in what follows. Thus the optical potential has only scalar and vector contributions, and the Dirac equation for the projectile has precisely the same form as in (7.2) and (7.3), with U^V replacing $g_v V^0$ and U^S replacing $(-g_s\phi)$:

$$h\mathcal{U}_0(\boldsymbol{x}) = \left\{-i\boldsymbol{\alpha}\cdot\boldsymbol{\nabla}+U^V(r; E)+\beta\left[M+U^S(r; E)\right]\right\}\mathcal{U}_0(\boldsymbol{x}) = E\,\mathcal{U}_0(\boldsymbol{x}). \tag{7.39}$$

Here E is the total nucleon–nucleus cm projectile energy, and M is its rest mass. In practice, one includes in U^V the Coulomb potential computed from the empirical nuclear charge density; other electromagnetic contributions arising from the proton anomalous magnetic moment are of similar size to the tensor term U^T and are neglected here.

To compute the optical potentials, one needs the Fourier transform

$$\int\frac{\mathrm{d}^3q}{(2\pi)^3}e^{i\boldsymbol{q}\cdot\boldsymbol{x}}\widetilde{f}(q) = \frac{g^2}{4\pi}\frac{\Lambda^2}{\Lambda^2-m^2}\left\{\frac{\Lambda^2}{\Lambda^2-m^2}\frac{e^{-mr}-e^{-\Lambda r}}{r} - \frac{\Lambda}{2}e^{-\Lambda r}\right\}, \tag{7.40}$$

where $\widetilde{f}(q)$ is one of the terms on the right-hand side of (7.28). Equation (7.40) holds for all meson exchanges except for the pseudoscalar meson with derivative coupling, in which case one requires

$$\int\frac{\mathrm{d}^3q}{(2\pi)^3}e^{i\boldsymbol{q}\cdot\boldsymbol{x}}\widetilde{f}(q)\frac{q^2}{4M^2}$$

$$= \frac{\Lambda^2}{4M^2}\frac{g^2}{4\pi}\frac{\Lambda^2}{\Lambda^2-m^2}\left\{\frac{m^2}{\Lambda^2-m^2}\frac{e^{-\Lambda r}-e^{-mr}}{r} + \frac{\Lambda}{2}e^{-\Lambda r}\right\}. \tag{7.41}$$

For laboratory energies near 200 MeV, it is necessary to correct the optical potentials for medium modifications from Pauli blocking. These can be approximated by performing a Dirac–Brueckner calculation [7.29–33] using a one-boson-exchange potential. The result is a "Pauli blocking factor" $a(E)$ that can be used to modify the potentials with a local-density approximation as follows:

$$U^L(r; E) \rightarrow \left[1 - a(E)\left(\frac{\rho_B(r)}{\rho_0}\right)^{2/3}\right]U^L(r; E). \tag{7.42}$$

Here $\rho_B(r)$ is the local baryon density of the target and $\rho_0 = 0.1934$ fm^{-3}. The approximate dependence on $\rho_B^{2/3}$ agrees with phase-space arguments based on isotropic scattering. For more details and the values of $a(E)$ used below, see Ref. [7.19].

To solve the scattering-state Dirac equation (7.39), we use a different approach from that used for the bound states of Sect. 7.2. The wave function is first separated into its upper and lower two-component wave functions, and the Dirac equation is rewritten as two coupled, first-order differential equations. The lower components are then eliminated, leading to a single second-order differential equation for the upper components. This equation contains local, spin-orbit, and nonlocal ("Darwin") potentials, but the nonlocality can be removed by rewriting the upper wave function as

$$\mathcal{A}^{1/2}(r; E)u(\boldsymbol{x}), \quad \mathcal{A}(r; E) \equiv 1 + \frac{U^S(r; E) - U^V(r; E)}{E + M}. \tag{7.43}$$

After some algebra, the equation for the new two-component function $u(\boldsymbol{x})$ becomes

$$\left(-\boldsymbol{\nabla}^2 + V_{\text{cent}} + V_{\text{so}}\,\boldsymbol{\sigma} \cdot \boldsymbol{L} + V_{\text{Darwin}}\right)u(\boldsymbol{x}) = (E^2 - M^2)u(\boldsymbol{x}), \tag{7.44}$$

where the energy-dependent optical potentials are

$$V_{\text{cent}}(r; E) \equiv 2MU^S + 2EU^V + (U^S)^2 - (U^V)^2, \tag{7.45}$$

$$V_{\text{so}}(r; E) \equiv -\frac{1}{r}\frac{B'}{B}, \tag{7.46}$$

$$V_{\text{Darwin}}(r; E) \equiv \frac{3}{4}\left(\frac{B'}{B}\right)^2 - \frac{1}{r}\frac{B'}{B} - \frac{1}{2}\frac{B''}{B}. \tag{7.47}$$

Here the prime denotes a radial derivative and $\mathcal{B}(r; E) \equiv (E + M)\mathcal{A}(r; E)$.

Since the upper (and lower) two-component Dirac wave functions are eigenstates of $\boldsymbol{\sigma} \cdot \boldsymbol{L}$, (7.44) can be recast in a form with only a second derivative. It can then be solved by the well-known Numerov method [7.34]. Note that $u(\boldsymbol{x})$ is *not* equal to the upper component wave function in the region of the potential, but since $A(r; E) \to 1$ as $r \to \infty$, u has the same asymptotic behavior as the wave function at large r. Thus the correct boundary conditions can be imposed by matching u to the form of a Coulomb scattering solution incident in the $\hat{z}$ direction [7.35]:

$$\psi(r) \underset{r \to \infty}{\propto} \left\{ e^{i[pz - \eta \ln 2pr \sin^2 \theta/2]}\left[1 - \frac{\eta^2}{2ipr \sin^2 \theta/2}\right]\right\}\chi_{\text{inc}}$$

$$+ \frac{e^{i[pr - \eta \ln 2pr]}}{r}[A(\theta) + B(\theta)\boldsymbol{\sigma} \cdot \hat{n}]\chi_{\text{inc}}, \tag{7.48}$$

where p is defined by $E = \sqrt{p^2 + M^2}$, χ_{inc} is a two-component Pauli spinor, θ is the cm scattering angle, $\hat{n}$ is the normal to the scattering plane, and $\eta \equiv Ze^2M/p^2$, with Z the nuclear charge. The scattering observables are then easily determined from the scattering amplitude as

$$\frac{d\sigma}{d\Omega} = |A(\theta)|^2 + |B(\theta)|^2, \quad A_y = \frac{2\text{Re}\left[A^*(\theta)B(\theta)\right]}{d\sigma/d\Omega},$$

$$Q = \frac{2\text{Im}\left[A(\theta)B^*(\theta)\right]}{d\sigma/d\Omega}. \tag{7.49}$$

7.5   Numerical Methods

The program TIMORA solves the self-consistent Dirac–Hartree equations for closed-shell nuclei. This is done by a straightforward iteration of (7.11) to (7.14), using (7.9) to compute the densities. The initial input consists of meson masses and couplings, guesses for the Dirac eigenvalues, and approximate meson potentials parametrized in the familiar Woods–Saxon form:

$$\phi(r) = \phi(0)\Big\{1 + e^{(r-R)/a}\Big\}^{-1} \tag{7.50}$$

and similarly for the other fields.

With this input, subroutine DIRAC solves (7.13) and (7.14) using a standard fourth-order Runge–Kutta algorithm [7.36]. One first integrates outward from small r to the chosen match radius r_m and then integrates inward from large r to r_m. The solutions are scaled so that G_a is continuous at r_m, and the wave function is then normalized according to (7.8). The discontinuity in F_a is used to adjust the eigenvalue according to

$$\delta E_a = -M G_a(r_m)\big[F_a(r_m^+) - F_a(r_m^-)\big], \tag{7.51}$$

where $F_a(r_m^\pm)$ are the values of the lower component as one approaches the match radius from either side. [Equation (7.51) can be derived by taking the expectation value of the Hamiltonian (7.3) with the approximate wave function.] This matching procedure is repeated until $|\delta E_a|$ is less than the input Dirac tolerance. The same procedure determines the wave functions for all the occupied states.

The wave functions are then inserted in (7.9) to generate the source terms for the meson field equations (7.11) and (7.12). The fields are computed in subroutine MESON by integrating over the Yukawa Green function, which incorporates the boundary conditions of exponential decay at large r and vanishing slope for the fields at the origin. (The Coulomb Green function is handled as a special case.) For example, for the scalar field, we use

$$\phi(r) = g_s \int_0^\infty r'^2\, dr'\, \rho_s(r') \frac{1}{m_s r r'} \sinh(m_s r_<) \exp(-m_s r_>) \tag{7.52}$$

$$= \frac{g_s}{2 m_s r}\Big\{ e^{-m_s r}\big(I_1(r) - I_2(0)\big) + e^{m_s r} I_2(r) \Big\}, \tag{7.53}$$

where

$$I_1(r) \equiv \int_0^r r'\, dr'\, \rho_s(r') e^{m_s r'}, \quad I_2(r) \equiv \int_r^\infty r'\, dr'\, \rho_s(r') e^{-m_s r'}. \tag{7.54}$$

The integrals I_i are evaluated with Simpson's rule using a step size that is half of the input grid spacing. This requires an interpolation (carried out in subroutine HALF), but the resulting quadrature is valid at both odd and even input grid points. Notice that this procedure evaluates the field at all N grid points with $O(N)$ operations.

Once the fields are determined, subroutine DIRAC computes a new set of wave functions and eigenvalues, which in turn produce a new set of meson fields. This procedure is iterated until all new eigenvalues differ from their values in the previous iteration by less than the Hartree tolerance. The convergent results are used in (7.9) to yield the Dirac–Hartree densities for the spherical nucleus.

With the densities in hand, the program FOLDER can be used to compute the coordinate-space scalar and vector optical potentials. The algorithm basically follows (7.34–36) with simple parametrizations for the NN scattering amplitude. There are two choices for the parametrization: (a) the relativistic Love–Franey (RLF) model discussed in Sect. 7.3 with its separation of direct and exchange amplitudes, and (b) a parametrization given by McNeil, Ray, and Wallace (MRW) which uses complex Gaussian functions and treats the full amplitude as a direct term [7.26]. The former is preferred for proton energies below $T_{\text{lab}} = 400\,\text{MeV}$, while parameter sets exist at higher energies for the latter choice.

If the RLF option is chosen, the t^L functions are given by (7.26–28,32). Recall also that the densities in (7.33) involve different sums over occupied states for pp and pn scattering. It is most convenient to shift variables from $\boldsymbol{x}' \to \boldsymbol{x}' + \boldsymbol{x}$, so that the t^L functions are not involved in the angular integration. After performing the ϕ' integration, we find

$$U^L(r; E) = -\frac{8\pi^2 i p}{M}\left[\int_0^\infty dr'\, t_D^L(r'; E)\int_{-1}^1 dw\, \rho^L(|\boldsymbol{x}' + \boldsymbol{x}|)\right.$$
$$\left. + \int_0^\infty dr'\, t_X^L(r'; E) j_0(pr')\int_{-1}^{+1} dw\, \rho^L(\boldsymbol{x}' + \boldsymbol{x}, \boldsymbol{x})\right], \quad (7.55)$$

where $w = \cos\theta'$, $|\boldsymbol{x}' + \boldsymbol{x}|^2 = r'^2 + r^2 + 2wrr'$, and

$$\left|\frac{\boldsymbol{x}' + 2\boldsymbol{x}}{2}\right|^2 = \frac{1}{4}(r'^2 + 4wrr' + 4r^2)\,.$$

This last relation is used in (7.36).

The integrals in (7.55) are evaluated by Gauss–Laguerre quadrature. Since the nuclear density is now needed at $|\boldsymbol{x}' + \boldsymbol{x}|$, the radial integration must go out to roughly *twice* the nuclear radius. About 32 Gauss points are sufficient for the r' and w integrals. As discussed above, for spherical nuclei only the scalar, vector, and tensor terms exist, and we keep only the first two, as the tensor term is small. If requested, the Pauli blocking factor is applied using (7.42).

If the MRW option is taken, the optical potential is evaluated somewhat differently. We first transform the densities $\rho^L(\boldsymbol{x})$ to momentum space, then multiply by $\mathcal{F}^L(q, E)$ and transform back, leading to

$$U^L(r; E) = -\frac{4\pi i p}{M}\int\frac{d^3 q}{(2\pi)^3}\,e^{i q \cdot x}\mathcal{F}^L(q, E)\int d^3 x'\, e^{-i q \cdot x'}\rho^L(r')\,. \quad (7.56)$$

At each proton energy E, the NN amplitudes $\mathcal{F}^L(q, E)$ are parametrized as

$$F^L(q, E) = \mathcal{F}_0^L(E) e^{-q^2 \beta^L(E)} , \tag{7.57}$$

with one set of parameters $\mathcal{F}_0^L$ and β^L for the pp amplitudes and another for pn [7.26]. Note that (7.56) is evaluated as a direct term only, and one adds the contributions from occupied proton and neutron states.

Equation (7.56) can be reduced to

$$U^L(r; E) = -\frac{4\pi i p}{M}$$
$$\left(\frac{2}{\pi} \right) \frac{1}{r} \int_0^\infty dq \, \sin qr \int_0^\infty dr' \, r' \sin qr' \mathcal{F}_0^L(E) e^{-q^2 \beta^L(E)} \rho^L(r') . \tag{7.58}$$

This integral is also evaluated by a double Gaussian sum, but more points are needed than in (7.55), typically 50 each for r' and q. The upper limit on r' is chosen at 1.2 times the original maximum grid point to make the resulting potential smooth for all r. [$U(r)$ is computed on the *same* grid as the original density; if one fails to extend the range of r', the results suffer from "Gibbs' oscillations".] A simple exponential decay is assumed for $\rho^L(r')$ to define it at radii beyond the original mesh. The integration on q is taken from 0.0 to 10.0 fm^{-1}.

The resulting optical potentials are used as input in the program HOOVER, which calculates observables for proton elastic scattering from Dirac optical potentials. HOOVER is not sophisticated and is intended to be used only for producing scattering observables from the output of FOLDER. It is designed to be as friendly and simple-minded as its namesake (with apologies to the New England Aquarium).

HOOVER calculates scattering observables by finding the large-r behavior for solutions of Schrödinger-like equations with central and spin–orbit potentials:

$$\left(-\nabla^2 + V_{\text{cent}}(r) + V_{\text{so}}(r) \boldsymbol{\sigma} \cdot \boldsymbol{L} \right) u(\boldsymbol{x}) = p^2 u(\boldsymbol{x}) . \tag{7.59}$$

As discussed above, proton scattering in the Dirac formalism can also be described by a Schrödinger-like equation [see (7.44)], so (7.59) is applicable with the appropriate substitutions. In particular, V_{cent} is now the sum of the central and Darwin terms of (7.45) and (7.47), plus the nuclear Coulomb potential, while V_{so} is given in (7.46). The Coulomb potential is calculated by assuming a uniform spherical charge distribution of radius $R = r_0 A^{1/3}$, where $r_0 \approx 1.25$ fm. Here $p^2 = E^2 - M^2$ as in (7.44), where E is the nucleon–nucleus cm energy of the projectile, and M is its rest mass.

A partial-wave expansion of (7.59) leads to the radial equations

$$\left\{ \frac{d^2}{dr^2} + p^2 - \left[V_{\text{cent}}(r) + \left(\begin{matrix} l \\ -l-1 \end{matrix} \right) V_{\text{so}}(r) \right] - \frac{l(l+1)}{r^2} \right\} u_l^\pm(pr) = 0 , \tag{7.60}$$

where the reduced radial functions $u_l^{\pm}(pr)$ play the same role as the Coulomb functions F_l and G_l in a problem with no nuclear potential. Starting with $u_l^{\pm}(0)=0$, $u_l^{\pm}$ is integrated to large r with the Numerov algorithm where it is matched to the appropriate linear combination of asymptotic functions via

$$u_l^{\pm}(pr) \sim F_l(pr) + C_l^{\pm}[G_l(pr) + iF_l(pr)] . \qquad (7.61)$$

F_l, G_l, and σ_l (see below) are computed from the asymptotic series given in Refs. [7.35,37]. Note that (7.61) is only approximate for the Dirac solutions, since it assumes that we can match the Dirac wave functions to nonrelativistic Coulomb wave functions at large radius; this neglects terms of $O(V_{\mathrm{coul}}/M)$.

The scattering amplitudes $A(\theta)$ and $B(\theta)$ are obtained from the relations

$$A(\theta) = f_c(\theta) + \frac{1}{p}\sum_l e^{2i\sigma_l}[(l+1)C_l^+ + lC_l^-]P_l^1(\cos\theta) , \qquad (7.62)$$

$$B(\theta) = \frac{i}{p}\sum_l e^{2i\sigma_l}[C_l^+ - C_l^-]P_l^1(\cos\theta) , \qquad (7.63)$$

where $f_c(\theta)$ is the Rutherford scattering amplitude,

$$f_c(\theta) = -\frac{\eta}{2p\sin^2\theta/2} e^{-i\eta\ln(\sin^2\theta/2)+2i\sigma_0} . \qquad (7.64)$$

Here the σ_l are the Coulomb phase shifts, calculated from

$$\sigma_l = \arg\Gamma(l+1+i\eta) , \qquad \eta = \frac{MZe^2}{p^2} , \qquad (7.65)$$

and $P_l^1(\cos\theta)$ are associated Legendre polynomials. These results follow from the representation in (7.48) [7.27], and the scattering observables can be determined from (7.49). To achieve reasonable accuracy at intermediate scattering angles, HOOVER uses double-precision arithmetic to evaluate the sums in (7.62) and (7.63).

7.6   The Codes

The programs TIMORA, FOLDER, and HOOVER are unadorned implementations of the techniques described above. The user first runs TIMORA to produce scalar and baryon densities for neutrons and protons. FOLDER processes the densities into Dirac scalar and vector optical potentials that serve as input to HOOVER, which adds a Coulomb potential and computes the scattering observables.

7.6.1 TIMORA

The input specifies the meson masses and couplings, initial guesses for the meson fields, characteristics of the occupied orbitals, and information on possible valence orbitals (outside the doubly magic core). The code contains essentially three parts: MAIN reads input and calculates derived quantities such as rms radii; subroutine DIRAC solves (7.13) and (7.14) as described above, sums the results to compute the baryon and scalar densities, and writes out results when the calculation has converged; subroutine MESON solves the meson field equations (7.11) and (7.12) by integrating over the Yukawa Green function and the densities computed by DIRAC.

The input is read from TIMORA.INP and consists of the following variables:

line	variable	type	explanation
1	XMM	single	nucleon mass (MeV)
	HH	single	radial grid spacing (fm)
	IMAX	integer	number of grid points in radial grid (≤ 600)
2	XMASS(1:3)	single	scalar, vector, and rho meson masses (MeV)
3	G2(1:4)	single	squared coupling constants $g_s^2, g_v^2, g_\rho^2, \alpha$
4	S0, V0	single	guesses for scalar and vector fields (MeV) at the origin
	XR	single	Woods–Saxon radius parameter (fm)
	A	single	Woods–Saxon surface thickness (fm)
5	CONVRG	single	Hartree convergence tolerance (MeV)
	DCONVR	single	Dirac convergence tolerance (MeV); should be $\approx 0.1 *$ $CONVRG$
	B0, A0	single	guesses for rho and Coulomb field strengths at origin for Woods–Saxon form
6, ...			input for Dirac orbitals; see text

Note that all of the boson fields are conventionally positive, and all are initially assumed to have a Woods–Saxon shape, as in (7.50). (This is an adequate starting point even for the Coulomb field.) Input lines 1 through 5 have free-field format.

Line 6 of the input file specifies the number ISTATE of occupied single-particle orbitals. The next ISTATE input lines each contain the characteristics of a single occupied orbital in the following order:

- The number of nucleons ($\leq 2j + 1$). The orbital need not be full.

- The angular-momentum variable κ.
- An initial guess for the eigenvalue in MeV. A negative value is interpreted as a guess, while a positive value sets the initial guess for the ith orbital in the list equal to the converged eigenvalue of orbital $i-1$. The input value for the deepest proton and neutron levels should therefore be negative. These eigenvalue inputs should be chosen with care, since they determine the number of nodes in the radial wave function.
- Isospin of the level (proton $= 1/2$, neutron $= -1/2$).
- An alphanumeric label for the orbital (format 2A3), such as $1s1/2$. The first character should be a blank.
- The match radius r_m in fm. One should avoid placing this near an expected node in the wave function.

After the input for the occupied orbitals, the next line contains the number of wave functions to be printed out. This is used after the computation has converged, so that the subsequent list of orbital characteristics can contain both occupied and valence orbitals; the only requirement is that every level be bound.

Once the input has been read in, the boson Green's functions are initialized, and the Hartree iterations are begun, as described in Sect. 7.5. The iterations proceed until all eigenvalues change by less than CONVRG. Various intermediate results are written to the screen and to TIMORA.LOG. When the calculation has converged, the densities are written to TIMORA.DEN with the format

$$r, \quad (\rho_B)_p, \quad (\rho_B)_n, \quad (\rho_s)_p, \quad (\rho_s)_n,$$

with r in fm and the densities in fm^{-3}. The log file contains the requested wave-function information as well as the final meson fields and derived quantities like the binding energy/nucleon and proton and neutron rms radii.

7.6.2   FOLDER

The user inputs parameters for the proton–nucleus collision and provides a table of proton and neutron vector (baryon) and scalar densities, such as the one generated by TIMORA. The output consists of a report of the progress of the calculation and a file containing the (complex) scalar and vector potentials on the same radial grid as the densities. The potentials can then be used by a scattering program, such as HOOVER.

The input for FOLDER is read from FOLDER.INP and is described in the following table:

line	variable	type	explanation
1	IFOLD	integer	= 1: folding with RLF parametrization = 2: folding with MRW parametrization
2	IPICPL	integer	= 1: pseudoscalar invariant in (7.17) = 2: pseudovector invariant from (7.19)
	IPBL	integer	= 0: no Pauli blocking = 1: Pauli blocking from (7.42)
3	ELAB	single	laboratory kinetic energy (MeV)
	AM2	single	target mass in amu
4	DR	single	radial grid spacing (fm) for input density and output potential
	NR	integer	number of grid points in radial grid (≤ 500)
	NRP	integer	number of Gauss nodes for r' integration
	NX	integer	number of Gauss nodes for $\cos\theta'$ integration

The variables should be entered in free-field format in the exact order given. (The input file also contains explanatory labels that must be retained for the file to be read correctly.) The RLF option is suitable for kinetic energies between 150 and 400 MeV. Parameter sets are included for energies 200, 300, and 400 MeV in RLF.DAT. The MRW option is suitable for energies ≥ 400 MeV. Parameter sets are included for energies 100, 200, ..., 1000 MeV in MRW.DAT. It should not be hard to add a new or improved set of parameters after examining this file and the SETUP subroutines. Suggested values for the number of Gauss nodes are NRP ≥ 32 and NX ≥ 32 for RLF, whereas NRP ≥ 50 and NX ≥ 50 are needed for MRW.

The user also supplies a density file, such as TIMORA.DEN, in which each line corresponds to a point on the radial grid with the input values for

$$r, \quad (\rho_B)_p, \quad (\rho_B)_n, \quad (\rho_s)_p, \quad (\rho_s)_n,$$

where r is in fm and the densities in fm^{-3}. As an example, the TIMORA.DEN file for ^{40}Ca, calculated on a grid of 0.04 fm with 300 total points might read

```
0.04    0.8962E-01    0.9476E-01    0.8473E-01    0.8961E-01
0.08    0.8956E-01    0.9469E-01    0.8466E-01    0.8954E-01
         ⋮             ⋮             ⋮             ⋮
```

| 11.96 | 0.2494E-07 | 0.4186E-08 | 0.2468E-07 | 0.4121E-08 |
| 12.00 | 0.2345E-07 | 0.3887E-08 | 0.2321E-07 | 0.3826E-08 |

Important: the input parameters DR and NR to FOLDER must correspond to the output grid of TIMORA.DEN.

After reading in the specifications from FOLDER.INP, some kinematic quantities are calculated in subroutine SETUP. Then, depending on the folding scheme chosen, RLFSET or MRWSET reads in the published NN parameters and chooses the set corresponding to the energy *closest* to the one requested by the user. (FOLDER *does not* interpolate the parameters for NN amplitudes; however, for the RLF option, the Pauli blocking parameters are interpolated.) Depending on the folding scheme chosen, RLFFLD evaluates (7.55) [or MRWFLD evaluates (7.58)] and returns the optical potential in the array POT. These values are then written to FOLDER.POT by the subroutine OUTPUT, which lists

$$ r, \quad \mathrm{Re}\,U^{S}, \quad \mathrm{Im}\,U^{S}, \quad \mathrm{Re}\,U^{V}, \quad \mathrm{Im}\,U^{V} $$

for each grid point, with r in fm and the potentials in MeV. A report on the calculation is given in FOLDER.LOG, where the kinematic quantities and parameters for the NN amplitudes are printed as a check on the calculation.

Other subroutines and functions are:

- RHOINT: Performs linear interpolation to get nuclear densities at any r.
- GAULEG: Sets up array of Gauss points and weights [7.36].
- FDIR: Calculates the direct NN amplitude in the RLF model, using (7.26) and (7.28).
- FEXC: Calculates the exchange NN amplitude in the RLF model, using (7.27) and (7.28).
- FFUN: Evaluates the right-hand side of (7.40).
- FFUNPV: Evaluates the right-hand side of (7.41).
- SBESJ0: Evaluates $j_0(r)$.
- SBESJ1: Evaluates $j_1(r)$.
- EXPOK: Keeps e^{-r} from underflowing for large r.

7.6.3   HOOVER

The user inputs parameters for the scattering variables, and relevant kinematical quantities are calculated in SETUP. Subroutine GETPOT reads the potentials from disk (see below) and stores them in a form convenient for the Numerov integration of (7.60).

As a first step, subroutine COULOMB calculates the σ_l (7.65) for all partial waves and returns the values of $F_l(pr)$ and $G_l(pr)$ at two neighboring grid points at large r. The Numerov integration is then done in INTUP, which produces two values of $u_l^{\pm}$ at neighboring grid points with large r. In XSECT, the quantities $A(\theta)$ and $B(\theta)$ are calculated from the matching conditions (7.62) and (7.63); the observables are then computed for each cm angle and written to the output file HOOVER.OBS with the format

$$\theta_{cm}, \qquad \frac{d\sigma}{d\Omega}, \qquad A_y, \qquad Q,$$

with θ_{cm} in degrees and $d\sigma/d\Omega$ in mb/sr. (A_y and Q are dimensionless.) Note that large-angle observables are very sensitive to the approximations; the methods used here should be accurate for momentum transfers $q \lesssim 3\,\mathrm{fm}^{-1}$.

Other subroutines are:

- COUVAL: Performs intermediate calculations for the Coulomb functions.
- COUWF: Completes the calculation of F_l and G_l.
- GETSIG: Computes σ_l in all partial waves.
- LEG: Computes Legendre and associated Legendre polynomials for given θ.

The input for HOOVER is read from HOOVER.INP and described in the table on the next page. The variables should be entered in free-field format in the exact order given. (The input file also contains explanatory labels that must be retained for the file to be read correctly.) IPOTOP specifies the method of input for the potential. The two choices are:

(i) IPOTOP = 1:   The relativistic optical potential is read from FOLDER.POT. This file has RMAX/DR lines of freely formatted numbers, in the order

$$r, \quad \mathrm{Re}\,U_S, \quad \mathrm{Im}\,U_S, \quad \mathrm{Re}\,U_V, \quad \mathrm{Im}\,U_V.$$

This is the form in which the current version of FOLDER gives its output. As an example, consider the optical potential file for 400 MeV proton scattering from ^{208}Pb calculated by FOLDER, using the RLF model with PV invariant and with the Pauli blocking prescription. The file FOLDER.POT contains the lines

```
0.040   -0.2583E+03   0.5388E+02   0.1824E+03   -0.6431E+02
0.080   -0.2583E+03   0.5388E+02   0.1824E+03   -0.6431E+02
          ⋮              ⋮             ⋮             ⋮
11.960  -0.1658E-01   0.1201E-01   0.3634E-01   -0.1275E-01
12.000  -0.1497E-01   0.1131E-01   0.3443E-01   -0.1201E-01
```

line	variable	type	explanation
1	THETA0	double	smallest cm angle for observables (degrees)
	THMAX	double	largest cm angle for observables (degrees)
	DTHETA	double	interval between angles (degrees)
2	ALMAX	double	number of partial waves required; 80 should be sufficient at energies ELAB $\lesssim 1000\,\mathrm{MeV}$ and momentum transfers $\lesssim 3\,\mathrm{fm}^{-1}$
	DR	double	radial grid spacing (fm) of input potential
	RMAX	double	largest radius (fm) of the input potential
3	ELAB	double	laboratory kinetic energy (MeV)
	AM1	double	projectile mass in amu
	Z1	double	projectile charge
	AM2	double	target mass in amu
	Z2	double	target charge
	RCOU0	double	reduced Coulomb radius r_0 (fm) for uniform charge distribution
4	IPOTOP	integer	= 0: use Woods–Saxon parameters = 1: use potentials from FOLDER

(ii) IPOTOP = 0: Here the scalar and vector potentials are taken from a Woods–Saxon parametrization:

$$\mathrm{Re}\,U_S(r) = \frac{U_S^0}{\left[1 + \exp\left((r - R_S^R)/a_S^R\right)\right]},$$

$$\mathrm{Im}\,U_S(r) = \frac{W_S^0}{\left[1 + \exp\left((r - R_S^I)/a_S^I\right)\right]},$$

and similarly for the vector potential $(S \to V)$. In this case, four lines must be added to end of the input file HOOVER.INP. These contain the parameters:

$$U_S^0, \quad R_S^R, \quad a_S^R;$$
$$W_S^0, \quad R_S^I, \quad a_S^I;$$
$$U_V^0, \quad R_V^R, \quad a_V^R;$$
$$W_V^0, \quad R_V^I, \quad a_V^I.$$

with U and W in MeV and R and a in fm. As all the input now comes in from `HOOVER.INP`, no input file of optical potentials is needed. An approximate set of parameters can be found in ref. [7.38] ; although these correspond to a slightly different parametrization than the one used here, they should be sufficiently accurate for generating qualitative results.

7.7   Hints and Things to Do

Calculate the scattering observables for 400 MeV proton scattering from ^{208}Pb and compare three different models for the constructing the RIA optical potentials:

(i) The RLF model with PV invariant, including Pauli blocking.
(ii) The RLF model with PV invariant, without Pauli blocking.
(iii) The MRW parametrization, which does not include Pauli blocking.

7.8   Technical Note by the Editors

The code "`HOOVER.FOR`" contains many instances of mixed-mode arithmetic between complex and double-precision types, which is not allowed by the FORTRAN 77 standard. Most compilers will accept this, but if the user experiences problems, it is relatively easy to remedy it either by (but with a consequent loss of accuracy) changing all double-precision quantities to real ones, or by explicitly using the SNGL function to convert all double-precision values to single precision only in the statements affected.

7.9   Acknowledgements

This work has been supported in part by DOE contract DE–FG02–87ER40365.

References

[7.1]   J. D. Walecka, Ann. Phys. (N.Y.) **83** (1974) 491
[7.2]   B. D. Serot and J. D. Walecka, Adv. Nucl. Phys. **16** (1986) 1
[7.3]   L. D. Miller and A. E. S. Green, Phys. Rev. **C5** (1972) 241
[7.4]   L. D. Miller, Phys. Rev. **C9** (1974) 537
[7.5]   R. Brockmann and W. Weise, Phys. Rev. **C16** (1977) 1282
[7.6]   L. N. Savushkin, Sov. J. Nucl. Phys. **30** (1979) 340
[7.7]   C. J. Horowitz and B. D. Serot, Nucl. Phys. **A368** (1981) 503
[7.8]   J. Boguta, Nucl. Phys. **A372** (1981) 386
[7.9]   C. J. Horowitz and B. D. Serot, Phys. Lett. **140B** (1984) 181
[7.10]   C. E. Price and G. E. Walker, Phys. Rev. **C36** (1987) 354
[7.11]   W. Pannert, P. Ring, and J. Boguta, Phys. Rev. Lett. **59** (1987) 2420

[7.12] R. J. Furnstahl, C. E. Price, and G. E. Walker, Phys. Rev. **C36** (1987) 2590

[7.13] J. A. McNeil, J. R. Shepard, and S. J. Wallace, Phys. Rev. Lett. **50** (1983) 1439

[7.14] J. R. Shepard, J. A. McNeil, and S. J. Wallace, Phys. Rev. Lett. **50** (1983) 1443

[7.15] B. C. Clark, S. Hama, R. L. Mercer, L. Ray, and B. D. Serot, Phys. Rev. Lett. **50** (1983) 1644

[7.16] B. C. Clark, S. Hama, R. L. Mercer, L. Ray, G. W. Hoffmann, and B. D. Serot, Phys. Rev. **C28** (1983) 1421

[7.17] B. C. Clark, S. Hama, and R. L. Mercer, in *The Interaction Between Medium Energy Nucleons in Nuclei*, edited by H. O. Meyer, AIP Conference Proceedings no. 97, (AIP, New York, 1983) p. 260

[7.18] C. J. Horowitz, Phys. Rev. **C31** (1985) 1340

[7.19] D. P. Murdock and C. J. Horowitz, Phys. Rev. **C35** (1987) 1442

[7.20] J. A. Tjon and S. J. Wallace, Phys. Rev. **C32** (1985) 1667

[7.21] J. D. Bjorken and S. D. Drell, *Relativistic Quantum Mechanics*, (McGraw-Hill, New York, 1964)

[7.22] A. R. Edmonds, *Angular Momentum in Quantum Mechanics*, 2nd edition (Princeton University Press, 1957)

[7.23] R. J. Perry, Phys. Lett. **182B** (1986) 269

[7.24] W. R. Fox, Nucl. Phys. **A495** (1989) 463

[7.25] P. G. Blunden, Phys. Rev. **C** (1990) in press

[7.26] J. A. McNeil, L. Ray, and S. J. Wallace, Phys. Rev. **C27** (1983) 2123

[7.27] D. P. Murdock, "Proton Scattering as a Probe of Relativity in Nuclei", Ph.D. Thesis, MIT, 1987

[7.28] F. A. Brieva and J. R. Rook, Nucl. Phys. **A291** (1977) 317

[7.29] C. J. Horowitz and B. D. Serot, Phys. Lett. **137B** (1984) 287

[7.30] R. Brockmann and R. Machleidt, Phys. Lett. **149B** (1984) 283

[7.31] R. Machleidt and R. Brockmann, Phys. Lett. **160B** (1985) 364

[7.32] B. ter Haar and R. Malfliet, Phys. Lett. **172B** (1986) 10; Phys. Rev. Lett. **56** (1986) 1237

[7.33] C. J. Horowitz and B. D. Serot, Nucl. Phys. **A464** (1987) 613; **A473** (1987) 760 (E)

[7.34] S. E. Koonin, *Computational Physics* (Benjamin, Reading, MA, 1986)

[7.35] I. E. McCarthy, *Introduction to Nuclear Theory* (Wiley, New York, 1968)

[7.36] W. H. Press, B. P. Flannery, S. A. Teukolsky, and W. T. Vetterling, *Numerical Recipes* (Cambridge University Press, 1986)

[7.37] M. Abramowitz and I. A. Stegun, *Handbook of Mathematical Functions with Formulas, Graphs, and Mathematical Tables* (Wiley, New York, 1964) Chap. 14

[7.38] A. M. Kobos, E. D. Cooper, J. I. Johansson, and H. S. Sherif, Nucl. Phys. **A445** (1985) 605

8. Three-Body Bound-State Calculations
W. Glöckle

8.1 Motivation

The He atom, with two electrons interacting both mutually and with the nucleus, historically played a significant role in demonstrating the validity of quantum mechanics in a system more complicated than the H atom. The bound states of three nucleons, ^{3}H and ^{3}He, are still not understood, since the nuclear interactions cannot yet be calculated rigorously from an underlying theory. Therefore, tests of similar basic nature still lie in the future. At present one uses either purely phenomenological forces or forces based on meson theory [8.1] and adjusts them to describe two-nucleon observables. The question is then whether these forces are also sufficient to describe three interacting nucleons or whether in addition three-nucleon forces are needed [8.2]. Since the three-body Schrödinger equation can be solved numerically in a precise manner, the three-nucleon system plays a very significant role in answering that question. One must also investigate, in addition to the bound state, three-nucleon scattering observables [8.3] and responses to external probes [8.4].

Three-atom systems are also of great interest. Here, a very exciting phenomenon for three identical bosons is the Efimov effect [8.5], in which the number of three-body bound states increases as $\ln|a|$ if the two-body binding energy goes to zero and the scattering length $|a|$ consequently tends to infinity. Such a case is very likely realized in the ^{4}He trimer [8.6].

In the future, three-quark systems will be of interest to test quantitatively models for the nucleon or, ultimately, QCD predictions.

8.2 Three-Body Faddeev Equations

The Faddeev equations [8.7] have been proven to be very useful and we shall concentrate on them in this chapter. Other approaches used to treat three-body systems are variational calculations [8.8], the use of hyperspherical harmonics [8.9], and more recently the Green's function Monte Carlo technique [8.10].

The Faddeev equations have been discussed extensively [8.11,12]. They transcribe the content of the Schrödinger equation (including boundary conditions) in a unique manner into a set of three coupled equations, whose kernel is connected after one iteration. For the bound state, they have the form

$$\psi_1 = G_0 T_1 \left(\psi_2 + \psi_3\right),$$
$$\psi_2 = G_0 T_2 \left(\psi_3 + \psi_1\right),$$

$$\psi_3 = G_0 T_3 (\psi_1 + \psi_2), \tag{8.1}$$

where

$$G_0 = \frac{1}{E - H_0} \tag{8.2}$$

is the free three-body propagator and T_i $(i = 1, 2, 3)$ are the two-body T-operators embedded in the three-particle space. They are connected to the pair forces V_i through the Lippmann–Schwinger equation (LSE)

$$T_i = V_i + V_i G_0 T_i . \tag{8.3}$$

Here we used the convenient "odd man out" notation

$$V_i \equiv V_{jk} \qquad j \neq i \neq k . \tag{8.4}$$

The wave function is given in terms of the three Faddeev components as

$$\Psi = \psi_1 + \psi_2 + \psi_3 . \tag{8.5}$$

For identical particles the three equations (8.1) reduce to a single one owing to permutation symmetries. This is most easily seen by going back to the definition of the Faddeev components

$$\psi_i \equiv G_0 V_i \Psi . \tag{8.6}$$

Let us apply the cyclical permutation $P_{12}P_{23}$ to ψ_1:

$$\begin{aligned}
P_{12}P_{23}\psi_1 &= P_{12}P_{23}G_0 V_1 \Psi = G_0 V_2 P_{12}P_{23}\Psi \\
&= G_0 V_2 \Psi = \psi_2 .
\end{aligned} \tag{8.7}$$

Similarly

$$P_{13}P_{23}\psi_1 = \psi_3 . \tag{8.8}$$

Consequently the first equation in (8.1) reads

$$\psi_1 = G_0 T_1 (P_{12}P_{23} + P_{13}P_{23}) \psi_1 \equiv G_0 T_1 P \psi_1 \tag{8.9}$$

and the wave function is

$$\Psi = (1 + P) \psi_1 . \tag{8.10}$$

The differential form of the Faddeev equations results by using the relation

$$G_0 T_1 = \frac{1}{E - H_0 - V_1} V_1 \tag{8.11}$$

in (8.9). This leads immediately to

$$(E - H_0 - V_1) \psi_1 = V_1 P \psi_1 . \tag{8.12}$$

This form is used in configuration-space calculations [8.13], while here we use the integral form (8.9) in momentum space.

8.3 Momentum-Space Representation

The relative motion of three particles is conveniently described by Jacobi
momenta

$$
\begin{aligned}
\boldsymbol{p}_i &\equiv \frac{1}{2}\left(\boldsymbol{k}_j - \boldsymbol{k}_k\right), \\
\boldsymbol{q}_i &\equiv \frac{2}{3}\left(\boldsymbol{k}_i - \frac{1}{2}\left(\boldsymbol{k}_j + \boldsymbol{k}_k\right)\right),
\end{aligned}
\tag{8.13}
$$

where $(ijk) = (123),\ (231),\ (312)$. Related to them are momentum states

$$
|\,\boldsymbol{p}_i\boldsymbol{q}_i\rangle \equiv |\,\boldsymbol{p}_i\rangle \otimes |\,\boldsymbol{q}_i\rangle
\tag{8.14}
$$

which are normalized as

$$
\langle\boldsymbol{p}_i\boldsymbol{q}_i\,|\,\boldsymbol{p}_i^{'}\,\boldsymbol{q}_i^{'}\rangle = \delta\left(\boldsymbol{p}_i - \boldsymbol{p}_i^{'}\right)\delta\left(\boldsymbol{q}_i - \boldsymbol{q}_i^{'}\right)
\tag{8.15}
$$

and which span the space of three-body states:

$$
\int \mathrm{d}\boldsymbol{p}_i\mathrm{d}\boldsymbol{q}_i\,|\,\boldsymbol{p}_i\boldsymbol{q}_i\rangle\,\langle\boldsymbol{p}_i\boldsymbol{q}_i\,| = 1_p \otimes 1_q .
\tag{8.16}
$$

Nuclear interactions are short-ranged and therefore act predominantly in
states of low orbital angular momenta. Thus we also introduce partial-
wave-projected states $|\,plm_\ell\rangle$ and $|\,q\lambda m_\lambda\rangle$. Both are defined by

$$
\langle\boldsymbol{k}^{'}\,|\,k\ell m\rangle = \frac{\delta\left(k - k'\right)}{kk'}Y_{\ell m}\left(\hat{k}^{'}\right) .
\tag{8.17}
$$

It follows that

$$
\langle k\ell m\,|\,k'\ell'm'\rangle = \frac{\delta\left(k - k'\right)}{kk'}\delta_{\ell\ell'}\delta_{mm'}
\tag{8.18}
$$

and

$$
\sum_{\ell m}\int \mathrm{d}k k^2\,|\,k\ell m\rangle\langle k\ell m\,| = 1_k .
\tag{8.19}
$$

We also form states of total orbital angular momentum L:

$$
|\,pq\,(\ell\lambda)\,LM\rangle = \sum_{m_\ell m_\lambda}\left(\ell\lambda L, m_\ell m_\lambda M\right)|\,p\ell m_\ell\rangle\,|\,q\lambda m_\lambda\rangle .
\tag{8.20}
$$

For notational simplicity we have dropped the indices that distinguish the
three choices of pairs of Jacobi variables. In the following we use the convenient notation

$$
|\,pq\,(\ell\lambda)\,LM\rangle_i ,
\tag{8.21}
$$

which means that p and ℓ refer to the relative motion of particles j and k,
while q and λ refer to the motion of particle i. The discrete set of quantum
numbers $(\ell\lambda)LM$ will be abbreviated in the following by α. It is obvious
that the basis states for each index obey the relations

$$\langle pq\alpha \mid p'q'\alpha' \rangle = \frac{\delta(p-p')}{pp'} \frac{\delta(q-q')}{qq'} \delta_{\alpha\alpha'} \tag{8.22}$$

and

$$\sum_\alpha \int \mathrm{d}p\, p^2 \int \mathrm{d}q\, q^2 \mid pq\alpha \rangle \langle pq\alpha \mid = 1_p \otimes 1_q \, . \tag{8.23}$$

In the following we will discuss in detail the simplest case of three identical bosons interacting by scalar pair interactions. Applications of the Faddeev equations to the dynamically richer three-nucleon system will be briefly indicated at the end of this section.

The partial-wave-projected momentum representation of a scalar pair interaction is defined by

$$\langle p\ell m \mid V \mid p'\ell'm' \rangle = \delta_{\ell\ell'} \delta_{mm'} \, v_\ell(p,p') \, . \tag{8.24}$$

This implies the representation in the three-particle basis (8.20):

$$_1\langle pq\alpha \mid V_1 \mid p'q'\alpha' \rangle_1 = \frac{\delta(q-q')}{qq'} \delta_{\alpha\alpha'} \, v_\ell(pp') \, . \tag{8.25}$$

The operator of kinetic energy H_0 is diagonal in the Jacobi momenta

$$H_0 = \frac{p^2}{m} + \frac{3}{4m} q^2 , \tag{8.26}$$

where m is the particle mass. Consequently

$$\langle pq\alpha \mid G_0(z) \mid p'q'\alpha' \rangle = \frac{\delta(p-p')}{pp'} \frac{\delta(q-q')}{qq'} \delta_{\alpha\alpha'} \frac{1}{z - \frac{p^2}{m} - \frac{3}{4m}q^2} \, . \tag{8.27}$$

The two-body T-operator in the three-particle space as given by the LSE (8.3) is clearly diagonal in the spectator quantum numbers q and λ but depends on q through the kinetic energy in G_0. The energy available to the interacting two-body subsystem is evidently $E - (3/4m)q^2$; consequently the T-operator is related to the proper two-body t-operator by

$$_1\langle pq\alpha \mid T_1 \mid p'q'\alpha' \rangle_1 = \frac{\delta(q-q')}{qq'} \delta_{\alpha\alpha'} t_\ell \left(p, p', E - \frac{3}{4m}q^2 \right) \tag{8.28}$$

and the partial-wave-projected two-body t-operator t_ℓ for an arbitrary off-shell energy z obeys

$$t_\ell(p,p',z) = v_\ell(p,p')$$
$$+ \int_0^\infty \mathrm{d}p''p''^2 v_\ell(p,p'') \frac{1}{z - \frac{p''^2}{m}} t_\ell(p'',p',z) . \tag{8.29}$$

We thus see that the study of the three-body system requires knowledge of the two-body off-shell t-matrices.

Finally we come to the permutation operator P defined in (8.9), which connects the three different ways of grouping three particles into a pair and a single particle. The evaluation of P in the basis (8.20) is a purely geometrical problem;

$$
\begin{aligned}
{}_1\langle pq\alpha \mid P_{12}P_{23} \mid p'q'\alpha'\rangle_1 &= {}_1\langle pq\alpha \mid p'q'\alpha'\rangle_2 \\
&= \int d\boldsymbol{p}''d\boldsymbol{q}'' \int d\boldsymbol{p}'''d\boldsymbol{q}''' \; {}_1\langle pq\alpha \mid \boldsymbol{p}''\boldsymbol{q}''\rangle_1 \\
&\quad {}_1\langle \boldsymbol{p}''\boldsymbol{q}'' \mid \boldsymbol{p}'''\boldsymbol{q}'''\rangle_2 \; {}_2\langle \boldsymbol{p}'''\boldsymbol{q}''' \mid p'q'\alpha'\rangle_2 \\
&= \int d\hat{p} \int d\hat{q} \int d\hat{p}' \int d\hat{q}' \; \mathcal{Y}_{\ell\lambda}^{LM*}(\hat{p}\hat{q}) \\
&\quad {}_1\langle p\hat{p}q\hat{q} \mid p'\hat{p}'q'\hat{q}'\rangle_2 \; \mathcal{Y}_{\ell'\lambda'}^{L'M'}(\hat{p}'\hat{q}') \; .
\end{aligned}
\tag{8.30}
$$

$\mathcal{Y}$ describes the spherical harmonics in ℓ and λ coupled to L and M. The remaining matrix element in (8.30) is a product of two three-dimensional δ-functions, which can be chosen in different ways. We choose

$$
{}_1\langle p\hat{p}q\hat{q} \mid p'\hat{p}'q'\hat{q}'\rangle_2 = \delta\left(p\hat{p} - \tfrac{1}{2}q\hat{q} - q'\hat{q}'\right) \delta\left(p'\hat{p}' + q\hat{q} + \tfrac{1}{2}q'\hat{q}'\right) , \tag{8.31}
$$

which will be seen below to be most useful [8.14]. It follows that

$$
\begin{aligned}
{}_1\langle pq\alpha \mid p'q'\alpha'\rangle_2 = \int d\hat{q} \int d\hat{q}' \frac{\delta\left(p- \mid \tfrac{1}{2}\boldsymbol{q} + \boldsymbol{q}' \mid\right)}{p^2} \\
\times \frac{\delta\left(p'- \mid \boldsymbol{q} + \tfrac{1}{2}\boldsymbol{q}' \mid\right)}{p'^2} \; \mathcal{Y}_{\ell\lambda}^{LM*}\left(\widehat{\tfrac{1}{2}\boldsymbol{q} + \boldsymbol{q}'}, \hat{q}\right) \; \mathcal{Y}_{\ell'\lambda'}^{L'M'}\left(-\widehat{\boldsymbol{q} - \tfrac{1}{2}\boldsymbol{q}'}, \hat{q}'\right) .
\end{aligned}
\tag{8.32}
$$

After expanding the spherical harmonics the three angular integrations can be performed analytically. For a detailed discussion see, for instance, [8.12]. Here we restrict ourselves to $\ell = \lambda = \ell' = \lambda' = L = L' = 0$, which finally leads to

$$
{}_1\langle pq\alpha \mid p'q'\alpha'\rangle_2 = \tfrac{1}{2}\int_{-1}^{1} dx \frac{\delta\left(p - \pi_1\right)}{p^2} \frac{\delta\left(p' - \pi_2\right)}{p'^2} \tag{8.33}
$$

with

$$
\pi_1 = \sqrt{\frac{1}{4}q^2 + q'^2 + qq'x} , \tag{8.34}
$$

$$
\pi_2 = \sqrt{q^2 + \frac{1}{4}q'^2 + qq'x}. \tag{8.35}
$$

The matrix element ${}_1\langle pq\alpha \mid P_{13}P_{23} \mid p'q'\alpha'\rangle_1$ can be reduced to the one for $P_{12}P_{23}$ by using

$$
P_{13}P_{23} = P_{23}P_{12}P_{23}P_{23} \tag{8.36}
$$

and applying P_{23} to the right and to the left. We get

$$_1\langle pq\alpha \mid P_{13}P_{23} \mid p'q'\alpha'\rangle_1 = (-)^{\ell+\ell'} \, _1\langle pq\alpha \mid P_{12}P_{23} \mid p'q'\alpha'\rangle_1 . \qquad (8.37)$$

Considering that for bosons both ℓ and ℓ' have to be even, we find that

$$_1\langle pq\alpha \mid P \mid p'q'\alpha'\rangle_1 = 2 \, _1\langle pq\alpha \mid p'q'\alpha'\rangle_2. \qquad (8.38)$$

Now we are prepared to write down the partial-wave-projected momentum space representation of the Faddeev equation (8.9):

$$\langle pq\alpha \mid \psi\rangle = \langle pq\alpha \mid G_0TP \ \psi\rangle$$
$$= \frac{1}{E - \frac{p^2}{m} - \frac{3}{4m}q^2}\langle pq\alpha \mid TP\psi\rangle. \qquad (8.39)$$

We insert the completeness relation (8.23) twice and use (8.28) to obtain

$$\langle pq\alpha \mid \psi\rangle = \frac{1}{E - \frac{p^2}{m} - \frac{3}{4m}q^2} \sum_{\alpha'} \int dp'p'^2 \int dq'q'^2 \langle pq\alpha \mid T \mid p'q'\alpha'\rangle$$
$$\times \sum_{\alpha''} \int dp''p''^2 \int dq''q''^2 \langle p'q'\alpha' \mid P \mid p''q''\alpha''\rangle\langle p''q''\alpha'' \mid \psi\rangle$$
$$= \frac{1}{E - \frac{p^2}{m} - \frac{3}{4m}q^2} \sum_{\alpha''} \int_0^\infty dp'p'^2 t_\ell\left(p, p', E - \frac{3}{4m}q^2\right)$$
$$\times \int dp''p''^2 \int dq''q''^2 \langle p'q\alpha \mid P \mid p''q''\alpha''\rangle\langle p''q''\alpha'' \mid \psi\rangle . \quad (8.40)$$

For the ground state, one has $L = 0$ and consequently $\ell = \lambda$. Let us consider now the simplest model case of pure s-wave interactions:

$$v_\ell = 0 \quad \text{for} \quad \ell \neq 0. \qquad (8.41)$$

Then from (8.40) it follows that only the Faddeev component

$$\psi(pq) \equiv \langle pq \, (\ell = 0, \lambda = 0) \ L = 0 \ M = 0 \mid \psi\rangle \qquad (8.42)$$

is nonvanishing. Using (8.33–35) and (8.38) we finally end up with

$$\psi(pq) = \frac{1}{E - \frac{p^2}{m} - \frac{3}{4m}q^2} \int_0^\infty dq'q'^2$$
$$\times \int_{-1}^1 dx \, t_0\left(p, \pi_1, E - \frac{3}{4m}q^2\right) \psi(\pi_2, q') . \qquad (8.43)$$

This homogenous integral equation in two variables has a nontrivial solution only at the correct binding energy E. It is for this single equation that we present numerical techniques in Sect. 8.4.

The realistic description of the three-nucleon problem is more difficult. The spins play a dynamical role. In jJ-coupling one combines the total two-body spin s with ℓ to the total two-body angular momentum j and similarily the spin $1/2$ with λ to the total angular momentum J of the third particle. Then j and J are coupled to the conserved total angular momentum $\mathcal{J}$. This leads to the following basis states:

$$| pq\alpha \rangle \equiv | pq\,(\ell s)j\left(\lambda\tfrac{1}{2}\right)\ J\,(jJ)\ \mathcal{J}M;\left(t\tfrac{1}{2}\right)TM_T\rangle. \tag{8.44}$$

Here, we have also added isospin quantum numbers and treat the three nucleons as identical in the framework of the generalized Pauli principle.

The generalization of (8.43) then yields [8.12] the following coupled set of integral equations:

$$\langle pq\alpha \mid \psi \rangle = \frac{1}{E - \frac{p^2}{m} - \frac{3}{4m}q^2} \sum_{\alpha'} \sum_{\alpha''} \int_0^\infty \mathrm{d}q' q'^2 \int_{-1}^1 \mathrm{d}x \tag{8.45}$$
$$\times\ \frac{t_{\alpha\alpha'}\left(p,\pi_1,E - \frac{3}{4m}q^2\right)}{\pi_1^{\ell'}}\ G_{\alpha'\alpha''}\left(q,q',x\right)\frac{\langle \pi_2\ q'\alpha''\mid\psi\rangle}{\pi_2^{\ell''}}.$$

One recognizes the two-body t-matrix $t_{\alpha\alpha'}$, which allows for tensor-force couplings, and the purely geometrical functions $G_{\alpha'\alpha''}\left(q,q',x\right)$, which result from the permutation operator P. For the triton with $\mathcal{J} = 1/2$ and positive parity, various numbers of discrete α-combinations, N_α, arise depending on the assumption of how many two-nucleon-force components are considered, as listed in Table 8.1. As an example for realistic three-nucleon calculations the last column shows the binding energies [8.15] based on the Argonne AV14 two-nucleon potential [8.16]. Various suggestions to explain the discrepancy with the experimental value of -8.48 MeV exist [8.17].

Table 8.1. The two-nucleon states in which V is nonvanishing together with N_α, the resulting number of discrete three-body states, and the corresponding three-nucleon binding energies.

	N_α	$E[MeV]$
$^1S_0, ^3S_1 - ^3D_1$	5	-7.44
$j \leq 2$	18	-7.57
$j \leq 4$	34	-7.67

8.4 Numerical Methods

8.4.1 Overview

The integral equation (8.43) is easily discretized in the variable q. One introduces, for instance, a cut-off value q_{max} and distributes properly Gauss–Legendre quadrature points over the intervals $0 \leq q \leq q_{\mathrm{max}}$. However, the situation is totally different in the variable p, which under the integral takes on the skew values π_2 of (8.35). Let N_q be the number of discrete q-points and N_x the corresponding number for the x-integration, then $N_q^2 \times N_x$ π_2-values occur. Typically we will have $N_q \approx 10$ and $N_x \approx 10$, which leads to a thousand π_2-values and therefore to too large a number of unknowns. Thus an interpolation appears to be unavoidable and adequate.

Before we discuss this point in detail, we note that the maximal value of π_2 is $3q_{\mathrm{max}}/2$, as is obvious from (8.35). This fact is of great importance in keeping the number of discretization points as low as possible and is a consequence of the way in which we have chosen the δ-functions in the evaluation of the permutation operator P in (8.31). The two-body subsystem is controlled by the variable p. The strong repulsive core in internucleon or interatomic forces leads to very large momentum components in the variable p, as we shall see in solving the two-body LSE. In other words the two-body t-matrix falls off very slowly. Obviously this carries over through the integral equation (8.43) to the Faddeev component $\psi(p,q)$. On the other hand the motion of the third particle is described through its distance to the center of mass of the two-body subsystem, which is the configuration-space coordinate conjugate to q. In that variable, the particle feels the two-body forces only averaged over a pair wave function, which is a less violent force. Consequently no large momentum components show up in the variable q. This will be born out in the numerical example. The value $3q_{\mathrm{max}}/2$ is therefore much lower than the typical cut-off value in the p-variable beyond which $\psi(p,q)$ can be neglected. Once the Faddeev equation (8.43) is solved in that smaller interval, $\psi(p,q)$ can be determined for all p-values just by quadrature using the integral equation again.

Let us now consider an interpolation in the form

$$ f(x) \approx \sum_k S_k(x) f(x_k), \tag{8.46} $$

where $S_k(x)$ are known functions and $\{x_k\}$ is a set of discrete grid points distributed over an interval in which f has to be determined. We apply this form to the p-variable in the Faddeev equation (8.43), choosing a suitable set of discrete p-points $\{p_k\}$. Then (8.43) can be approximated as

$$ \psi(pq) \approx \frac{1}{E - \frac{p^2}{m} - \frac{3}{4m}q^2} $$

$$ \times \int_0^\infty dq'q'^2 \sum_{k'} \int_{-1}^1 dx\, t\left(p, \pi_1, E - \frac{3}{4m}q^2\right) S_{k'}(\pi_2)\, \psi(p_{k'}, q'). \tag{8.47} $$

Though one could afford to calculate the two-body t-matrix at all π_1-values numerically it is "cheaper" to interpolate t as well. Finally we introduce Gaussian quadrature in the variable q' and get

$$ \psi(p_k, q_\ell) = \sum_{k'} \sum_{\ell'} \left\{ \frac{1}{E - \frac{p_k^2}{m} - \frac{3}{4m}q_\ell^2}\, w_{\ell'}\, q_{\ell'}^2 \right. $$

$$ \left. \times \sum_m t\left(p_k, p_m, E - \frac{3}{4m}q_\ell^2\right) \int_{-1}^1 dx\, S_m(\pi_1) S_{k'}(\pi_2) \right\} \psi(p_{k'}\, q_{\ell'}) \tag{8.48} $$

with

$$\pi_1 = \sqrt{\frac{1}{4}q_\ell^2 + q_{\ell'}^2 + q_\ell q_{\ell'} x} \ ; \tag{8.49}$$

$$\pi_2 = \sqrt{q_\ell^2 + \frac{1}{4}q_{\ell'}^2 + q_\ell q_{\ell'} x} \ . \tag{8.50}$$

This is a closed set of homogeneous algebraic equations with the matrix kernel as given inside the curly brackets. The remaining x-integral will also be performed numerically by Gaussian quadrature. If N_p is the number of grid points $\{p_k\}$ for the interpolation, $N \equiv N_p \times N_q$ is the number of unknowns and the kernel is a $N \times N$ matrix.

8.4.2 Details

The Spline Interpolation

A very useful realization of the interpolation (8.46) is by cubic splines [8.18]. These are piecewise cubic polynomials defined on the intervals between the grid points x_k and joined to the adjacent intervals such that the second derivative is continuous. The spline elements $S_k(x)$ can be determined by recurrence relations, as described in detail in [8.19].

The Two-Body Potential

For a local potential $V(r)$ the momentum representation in a state of angular momentum ℓ can be found in different ways. One is to use

$$\begin{aligned}
\langle \boldsymbol{p} \mid V \mid \boldsymbol{p}' \rangle &= \frac{1}{(2\pi)^3} \int \mathrm{d}\boldsymbol{r} \ e^{\mathrm{i}\left(\boldsymbol{p}-\boldsymbol{p}'\right)\cdot\boldsymbol{r}} \ V(r) \\
&= \frac{4\pi}{(2\pi)^3} \int_0^\infty \mathrm{d}r r^2 j_0\left(\mid \boldsymbol{p}-\boldsymbol{p}' \mid r\right) V(r) \\
&\equiv \tilde{V}\left(\mid \boldsymbol{p}-\boldsymbol{p}' \mid\right) .
\end{aligned} \tag{8.51}$$

Then

$$\begin{aligned}
v_\ell(p,p') &= \langle p\ell m \mid V \mid p'\ell m \rangle = \int \mathrm{d}\boldsymbol{p}'' \int \mathrm{d}\boldsymbol{p}''' \\
&\times \langle p\ell m \mid \boldsymbol{p}'' \rangle \tilde{V}\left(\mid \boldsymbol{p}'' - \boldsymbol{p}''' \mid\right) \langle \boldsymbol{p}''' \mid p'\ell m \rangle \\
&= \int \mathrm{d}\hat{p} \int \mathrm{d}\hat{p}' \ Y_{\ell m}^*(\hat{p}) \ \tilde{V}\left(\mid \boldsymbol{p}-\boldsymbol{p}' \mid\right) Y_{\ell m}(\hat{p}') .
\end{aligned} \tag{8.52}$$

Using the identity

$$\sum_m Y_{\ell m}^*(\hat{p}) Y_{\ell m}(\hat{p}') = \frac{2\ell+1}{4\pi} P_\ell(\hat{p}\cdot\hat{p}') \tag{8.53}$$

we get

$$v_\ell(p,p') = 2\pi \int_{-1}^{1} dx\, P_\ell(x)\; \tilde{V}\left(\sqrt{p^2+p'^2-2pp'x}\right). \tag{8.54}$$

A second way is to perform the transition between configuration and momentum space for a fixed ℓ:

$$\begin{aligned}
v_\ell(p,p') &= \int d\hat{p}\, Y_{\ell m}^*(\hat{p})\, \frac{1}{(2\pi)^{3/2}} \int d\mathbf{r}\, e^{i\mathbf{p}\cdot\mathbf{r}}\; V(r)\\
&\quad \times \int d\hat{p}'\, \frac{1}{(2\pi)^{3/2}} e^{-i\mathbf{p}'\cdot\mathbf{r}} Y_{\ell m}(\hat{p}')\\
&= \frac{2}{\pi} \int_0^\infty dr\, r^2 j_\ell(pr)\, V(r)\, j_\ell(p'r).
\end{aligned} \tag{8.55}$$

In the case of a Yukawa potential

$$V(r) = V_0 e^{-\mu r}/r, \tag{8.56}$$

which we shall use in the model calculations, both formulae (8.54) and (8.55) can be evaluated analytically with the result

$$v_\ell(p,p') = \frac{V_0}{\pi}\frac{1}{pp'}\, Q_\ell(z) \tag{8.57}$$

and

$$z = \frac{p^2+p'^2+\mu^2}{2pp'}. \tag{8.58}$$

Specifically for $\ell = 0$ one gets

$$v_0(p,p') = \frac{V_0}{\pi}\frac{1}{pp'}\frac{1}{2}\ln\frac{(p+p')^2+\mu^2}{(p-p')^2+\mu^2} \tag{8.59}$$

The Two-Body t-Matrix

The potential $v_\ell(p,p')$ enters into the driving term and the kernel of the LSE (8.29) for t_ℓ. The p''-integral in (8.29) can be conveniently discretized by N_p Gaussian quadrature points and we get in obvious notation

$$\begin{aligned}
t_\ell(p_k,p',z) &= v_\ell(p_k,p')\\
&\quad + \sum_{k'} w_{k'}' p_{k'}'^2 v_\ell(p_k,p_{k'})\frac{1}{z-p_{k'}'^2/m}\, t_\ell(p_{k'}',p',z).
\end{aligned} \tag{8.60}$$

For each p' and energy z this is a closed set of inhomogeneous algebraic equations with a matrix kernel of dimension $N_p \times N_p$. For a three-body bound state the two-body energies are

$$z = E - \frac{3}{4m}q^2 < 0, \tag{8.61}$$

and therefore the denominating part in (8.60) is nonsingular.

The Three-Body Eigenvalue Problem

The homogeneous Faddeev equation (8.43) can be regarded as an eigenvalue problem

$$\eta_\mu \psi_\mu = K \psi_\mu \qquad (8.62)$$

with $\eta_\mu = 1$. Since the square of the Faddeev kernel is connected and therefore has a discrete spectrum, the same applies to K. The point of accumulation of the set $\{\eta_\mu\}$ is zero. Consequently there is an eigenvalue $\eta_{\max}$, which is largest in magnitude. A simple way to find $\eta_{\max}$ is the simple vector iteration method [8.18], which was popularized in the context of three-body Faddeev equations by Malfliet and Tjon [8.20]. Choosing an arbitrary state ϕ one generates

$$\phi_n \equiv K^n \phi , \qquad n > 0, \qquad (8.63)$$

and forms the ratios

$$R_n = \phi_{n+1}/\phi_n. \qquad (8.64)$$

Decomposing ϕ into the set of eigenstates ψ_μ we clearly see that

$$\lim_{n\to\infty} R_n = \eta_{\max} \qquad (8.65)$$

and that ϕ_n tends towards the corresponding eigenstate.

For a purely attractive interaction all the eigenvalues η_μ are positive. If E is the ground-state energy, then $\eta_{\max}$ is the physical eigenvalue 1. This can easily be seen. Assume that $\eta_{\max} > 1$; then the kernel $K/\eta_{\max}$ with a weaker two-body t-matrix $t/\eta_{\max}$ would have a bound state at the energy E contrary to the assumption that E is the lowest energy related to t.

If V has attractive and repulsive parts, negative eigenvalues will also occur and it may be that $\eta_{\max} < -1$. Then the Malfliet–Tjon (MT) method converges towards that unphysical eigenvalue. A simple algorithm to overcome this problem is given in [8.14]. First one determines $\eta_{\max}$ by the MT method. An approximate value $\eta_{\max}^{\mathrm{appr}}$ is sufficient. Then we define

$$\phi' = (K - \eta_{\max}^{\mathrm{appr}}) \phi. \qquad (8.66)$$

Decomposing ϕ into the set of eigenstates ψ_μ, one sees that the coefficient of the eigenstate belonging to $\eta_{\max}$ is proportinal to $(\eta_{\max} - \eta_{\max}^{\mathrm{appr}})$. If necessary, the step from ϕ to ϕ' can be repeated in order to suppress the component belonging to $\eta_{\max}$ even more. The MT method applied to ϕ' will converge towards the eigenvalue which comes next in magnitude.

An even simpler algorithm is to replace the eigenvalue problem (8.62) by

$$(K - \eta_{\max}^{\mathrm{appr}}) \psi_\mu = (\eta_\mu - \eta_{\max}^{\mathrm{appr}}) \psi_\mu . \qquad (8.67)$$

Then η_{max} is mapped onto approximately zero and the physical eigenvalue 1 onto $(1 - \eta_{max}^{appr})$, which is the largest one in magnitude. Then the MT method can be directly applied to (8.67).

In the numerical examples given below, these modifications turn out to be unnecessary. Since the energy eigenvalue E is not known, one starts with an estimated energy and determines the corresponding physical eigenvalue η. Then the energy E is varied such that $\eta(E)$ approaches the value 1. This is a simple routine searching for the zero of

$$\eta(E) - 1 = 0. \tag{8.68}$$

8.5 Model Calculations

We choose a model of three nucleons interacting via a pure s-wave potential. The interaction is averaged over singlet and triplet states and has been given in [8.20]. It is of the form

$$V(r) = -V_A \frac{e^{-\mu_A r}}{r} + V_R \frac{e^{-\mu_R r}}{r}. \tag{8.69}$$

The parameters are displayed in Table 8.2 together with the two-body binding energy ϵ.

Table 8.2. The potential parameters of (8.69) and the two-body binding energy.

V_A [MeVfm]	V_R [MeVfm]	μ_A [fm^{-1}]	μ_R [fm^{-1}]	ϵ [MeV]
570.316	1438.4812	1.55	3.11	−0.35

In the three-body ground state of our model the space part is totally symmetric under exchange of two particles and the antisymmetric spin–isospin part of the wave function can be factored out. Therefore one has effectively the case of three identical bosons. The momentum-space potentials $v_0(p, p')$ have the form (8.59).

For the solution of the two-body LSE (8.60) we use Gaussian–Legendre quadrature points $\{p_k\}$ of the following type

$$p = \frac{(2ac - ab - bc)(1 + x) + 2a(b - c)}{(a - 2b + c)\,x\, + a - c} \qquad (-1 \le x \le 1). \tag{8.70}$$

Half of the n quadrature points lies in the p-interval (a, b) and the other half between b and c. The weights $\{w_k\}$ can be read off from

$$dp = \frac{2(b - c)(a - c)(b - a)}{((a - 2b + c)\,x\, + a - c)^2}\, dx. \tag{8.71}$$

The total p-interval necessary for solving (8.60) is split into a first part, $[0, p_{max}]$, which enters in solving the Faddeev equation (8.48), and a remaining part $[p_{max}, p_{cut}]$. Here $p_{max} = 3/2\, q_{max}$ (in fact we add a small quantity to $3/2\, q_{max}$ in order to allow for a safe interpolation) and p_{cut} is the cut-off value for the p''-integral in (8.29).

We choose the same distribution for the N_q q-points, where now $a = 0$ and $c = q_{max}$. For a given starting energy E the LSE (8.60) is to be solved at the N_q energies $z_k = E - 3q_k^2/4m$.

It is convenient to choose the set of grid points for the spline interpolation to be identical to the quadrature points in the first p-interval. Therefore the p'-values in (8.60) are identified with that first group of points. Let that number of points be N_p. Then for each energy z the inhomogeneous set (8.60) has N_p different driving terms:

$$\sum_{k''} \left(\delta_{kk''} - \frac{w_{k''} p_{k''}^2}{z - p_{k''}^2/m} v_0\left(p_k, p_{k''}\right) \right) t\left(p_{k''}, p_{k'}, z\right) = v_0\left(p_k, p_{k'}\right). \qquad (8.72)$$

This is a standard algebraic problem and is solved by the routine F04AEF of the NAG library.

The spline elements $S_k(p)$ are generated in two steps. A first subroutine (SPREP) prepares coefficients, which requires only the knowledge of the set of grid points $\{p_k\}$. A second subroutine (SELEM) calculates the N_p spline elements $S_k(p)$ for a given p-value.

Now we are ready to build up the Faddeev kernel in (8.48). Depending on the storage available the matrix kernel can be calculated once or built up out of prepared parts at each iteration within the MT method. We choose the second option. The structure of the kernel in (8.48) is very transparent and can be recovered easily in the code.

In Table 8.3 we display a set of "safe" discretization parameters. Here N_p^{tot} is the total number of Gaussian-quadrature p-points used for solving (8.72) out of which only the first N_p p-points are needed for solving (8.48). For the spline interpolation it is convenient to include $p = 0$ as the first point.

Table 8.3. Discretization parameters for (8.48) and (8.72).

N_p	p_{max}	p_{cut}	N_p^{tot}	N_q	q_{max}	N_x
19	6.3	50.0	39	10	4.0	10

The ratios R_n of (8.64) depend on p and q for small n and will become independent of these variables only for asymptotic values of n. Starting with the initial state $\phi(pq) \equiv 1$ we show in Table 8.4 the ratios R_n for the fixed p and q values $p = 0.4592$ fm^{-1}, $q = 0.7922$ fm^{-1}, and the maximal

Table 8.4. The minimal and maximal ratios $R_n^{(\pm)}$ of (8.64) together with the ratio $\bar{R}_n$ at the fixed values $p = 0.4592$ fm^{-1}, $q = 0.7922$ fm^{-1} as a function of n. The initial state is $\phi \equiv 1$.

n	$R_n^{(-)}$	$\bar{R}_n$	$R_n^{(+)}$
2	-0.422859E$+01$	0.552505E$+00$	0.165611E$+01$
3	-0.340999E$+01$	0.114284E$+01$	0.416641E$+02$
4	-0.146104E$+03$	0.933817E$+00$	0.248889E$+02$
5	-0.148619E$+02$	0.103320E$+01$	0.222942E$+02$
6	-0.485679E$+01$	0.998753E$+00$	0.292972E$+01$
7	-0.389955E$+01$	0.101066E$+01$	0.169167E$+03$
8	-0.377271E$+02$	0.100626E$+01$	0.290163E$+03$
9	-0.473449E$+02$	0.100782E$+01$	0.321938E$+02$
10	-0.592942E$+00$	0.100726E$+01$	0.652134E$+01$
11	0.324155E-01	0.100746E$+01$	0.133087E$+01$
12	0.919647E$+00$	0.100739E$+01$	0.118705E$+02$
13	0.677033E$+00$	0.100741E$+01$	0.104186E$+01$
14	0.995467E$+00$	0.100740E$+01$	0.118361E$+01$
15	0.953650E$+00$	0.100741E$+01$	0.101174E$+01$
16	0.100586E$+01$	0.100741E$+01$	0.102776E$+01$
17	0.100026E$+01$	0.100741E$+01$	0.100796E$+01$
18	0.100721E$+01$	0.100741E$+01$	0.100999E$+01$
19	0.100648E$+01$	0.100741E$+01$	0.100748E$+01$

deviations $R_n^{(\pm)}$ in both directions by going through all p- and q-values. The energy is $E = -7.4$ MeV. We see that a sufficiently perfect independence of p and q has not yet been reached within 19 iterations. The situation improves considerably if at a second energy the initial state ϕ is chosen to be the final one of the iteration procedure at the first energy. This is illustrated in Table 8.5. Table 8.6 demonstrates the energy search.

Including the tests with respect to variations of the discretization parameters we end up with a binding energy of $E = -7.54$ MeV.

Having reached the asymptotic value of R_n we have also found the unnormalized Faddeev component

$$\phi_n(p,q) \xrightarrow{n\to\infty} \psi(p,q). \tag{8.73}$$

According to (8.10) the total wave function is given by

$$\Psi(p,q) = \frac{1}{4\pi}\left(\psi(p,q) + \psi\left(|\frac{1}{2}p + \frac{3}{4}q|, |p - \frac{1}{2}q|\right)\right.$$
$$\left. + \psi\left(|\frac{1}{2}p - \frac{3}{4}q|, |p + \frac{1}{2}q|\right)\right). \tag{8.74}$$

Table 8.5. Same as in Table 8.4, but for the energy -7.5 MeV and using a starting vector from the end of the previous iteration at $E = -7.4$ MeV.

n	$R_n^{(-)}$	$\bar{R}_n$	$R_n^{(+)}$
2	0.998718	1.00313	1.02435
3	1.00005	1.00223	1.03643
4	1.00149	1.00212	1.00357
5	1.00181	1.00209	1.00588
6	1.00184	1.00208	1.00221
7	1.00205	1.00208	1.00250
8	1.00202	1.00208	1.00209
9	1.00207	1.00208	1.00212
10	1.00207	1.00208	1.00208

Table 8.6. Illustration of the energy search. The physical eigenvalue $\eta(E)$ of (8.62) as a function of E.

E [MeV]	$\eta(E)$
-7.4	1.0074
-7.5	1.0021
-7.6	0.99683
-7.5396	0.99999
-7.5394	1.0000

We have used the linear connection between the different sets of Jacobi momenta

$$p_2 = -\tfrac{1}{2}p_1 - \tfrac{3}{4}q_1 , \qquad p_3 = -\tfrac{1}{2}p_1 + \tfrac{3}{4}q_1 ,$$
$$q_2 = p_1 - \tfrac{1}{2}q_1 , \qquad q_3 = -p_1 - \tfrac{1}{2}q_1 . \qquad (8.75)$$

The state under consideration has $L = 0$ and thus depends only on p, q, and $x \equiv \hat{p} \cdot \hat{q}$. As mentioned in Sect. 8.4.1 the Faddeev component $\psi(p,q)$ leaks out in p much further than $3/2\, q_{max}$, up to which it has been determined. We can evaluate $\psi(p,q)$ for $p > 3q_{max}/2$ using the Faddeev equation (8.43) again; the kernel has to be built up choosing the desired p-values in the two-body t-matrix.

The two permuted Faddeev components for Ψ can be expanded into Legendre polynomials:

$$\Psi(p,q) = \frac{1}{4\pi} \sum_\ell \psi_\ell(p,q) P_\ell(x) (2\ell+1) \qquad (8.76)$$

with

$$\psi_\ell(p,q) = \int_{-1}^{1} dx P_\ell(x)\, \psi\left(\sqrt{\frac{1}{4}p^2 + \frac{9}{16}q^2 + \frac{3}{4}pqx}, \sqrt{p^2 + \frac{1}{4}q^2 - pqx}\right)$$
$$+ \delta_{\ell 0}\psi(p,q). \tag{8.77}$$

The evaluation of the x-integral requires ψ at skew values, which can be easily interpolated in terms of $\psi(p_k, q_\ell)$ using the given spline routine. Finally the normalization condition is

$$\int dp \int dq\, \Psi^2(p,q) = \sum_\ell (2\ell+1) \int dp p^2 \int dq q^2\, \psi_\ell^2(p,q) = 1. \tag{8.78}$$

One can also normalize Ψ directly without introducing the partial wave decomposition:

$$\langle \Psi \mid \Psi \rangle = \langle (1+P)\psi \mid (1+P)\psi \rangle \frac{1}{(4\pi)^2}$$
$$= \frac{3}{(4\pi)^2} \langle \psi \mid (1+P)\psi \rangle = 3\left[\int dp p^2 \int dq q^2 \psi^2(pq)\right.$$
$$+ 2 \int dp p^2 \int dq q^2 \int_{-1}^{1} dx \psi(pq) \tag{8.79}$$
$$\times \left. \psi\left(\sqrt{\frac{1}{4}p^2 + \frac{9}{16}q^2 + \frac{3}{4}pqx}, \sqrt{p^2 + \frac{1}{4}q^2 - pqx}\right)\right] = 1.$$

We leave the construction of the normalized wave function to the reader.

8.6 Exercises

The number of the various discretization points should be modified in order to get a feeling for the numerical stability.

Building up the Faddeev kernel only once for each energy instead of reconstructing it at each iteration will decrease the CPU time by about a factor of 3. Even more can be gained by vectorization.

The nucleon–nucleon interaction used has an attractive as well as a repulsive part. A similar interaction is the Malfliet–Tjon potential III (see Ref [8.20]) which gives a deuteron binding energy of -2.23 MeV. What would be the corresponding three-nucleon binding energy? The Malfliet–Tjon potential IV is purely attractive with the same deuteron binding of -2.23 MeV. Using this potential one will experience a dramatic increase in the three-nucleon binding energy, demonstrating the importance of the repulsion.

The total wave function is necessary, for instance, for the evaluation of the momentum distribution in the three-nucleon bound state. The normalized wave function can be obtained easily following the description given at the end of Sect. 8.5.

8.7   Technical Note

The following is a brief description of the NAG subroutines which are called by the program. The program D01BCF returns the weights and abscissae for a Gaussian quadrature, while subroutine F04AEF solves the algebraic equations (8.72). A more detailed description of these programs is found in the *NAG FORTRAN Library Manual*.

F04AEF

Purpose:		Calculates the solution of a set of real linear equations with multiple right-hand sides, AX=B, by Crout's factorization method.
Usage:		CALL F04AEF (A,IA,B,IB,N,M,C,IC,WK,AA,IAA, BB,IBB,IFAIL)
Arguments:		
A	–	real array of dimension (IA,p), where p ≥ N; A contains the element of the real matrix (input)
IA	–	first dimension of array A, IA ≥ N (input)
B	–	real array of dimension (IB,q), where q ≥ M; B contains the elements of the M right-hand sides (input)
IB	–	first dimension of array B, IB ≥ N (input)
N	–	specifies the order of matrix A (input)
M	–	specifies the number of right-hand sides (input)
C	–	real array of dimension (IC,r), where r ≥ M; C contains the M solution vectors (output)
IB	–	first dimension of array C, IC ≥ N (input)
WK	–	workspace array of dimension at least N
AA	–	real array of dimension (IAA,s) where s ≥ N; AA contains the Crout factorization (output)
IAA	–	first dimension of array AA, IAA ≥ N (input)
BB	–	real array of dimension (IBB,t) where t ≥ M; MM contains the M residual vectors (output)
IBB	–	first dimension of array BB, IBB ≥ N (input).
IFAIL	–	error indicator; before entry IFAIL must be assigned a value (IFAIL=0) (input).

D01BCF

Purpose:	Returns the weights and abscissae for a Gaussian quadrature rule with a specified number of abscissae.
Usage:	CALL D01BCF (IT,A,B,C,D,N,WEIGHT,ABSCIS,IFAIL)
Arguments:	
IT	– specifies the rule type, which is Gauss–Legendre (IT=0) in the present case (input)
A	– lower integral boundary (input)
B	– upper integral boundary (input)
C	– not needed presently (C=0) (input)
D	– not needed presently (D=0) (input)
N	– specifies the number of weights and abscissae to be returned (input)
WEIGHT	– array of dimension N containing the N weights (output)
ABSCIS	– array of dimension N containing the N abscissae (output)
IFAIL	– error indicator; before entry IFAIL must be assigned a value (IFAIL=0) (input)

References

[8.1] R. Machleidt, Adv. Nucl. Phys. **19** (1989) 189

[8.2] J.L. Friar, B.F. Gibson, and G.L. Payne, Ann. Rev. Nucl. Part. Sci. **34** (1984) 403; P.U. Sauer, Prog. Part. Nucl. Phys. **16** (1986) 35

[8.3] W. Glöckle, H. Witala, and Th. Cornelius, 12th International Conference on few-body problems in Physics, Vancouver, July 1989

[8.4] V.R. Pandharipande and R. Schiavilla, 12th International Conference on few-body problems in Physics, Vancouver, July 1989

[8.5] V. Efimov, Phys. Lett. **33B** (1970) 563; Soviet J. Nucl. Phys. **12** (1971) 589; Nucl. Phys. **A210** (1973) 157

[8.6] Th. Cornelius and W. Glöckle, J. Chem. Phys. **85** (1986) 3906

[8.7] L.D. Faddeev, Sov. Phys. JETP **12** (1961) 1014

[8.8] L.M. Delves, Adv. Nucl. Phys. **5** (1972) 1; R.B. Wiringa, Few-Body Systems Suppl **1** (1986) 130

[8.9] M. Fabre de la Ripelle, Lecture Notes in Physics **273** (1987) 283

[8.10] K.E. Schmidt, Lecture Notes in Physics **273** (1987) 363

[8.11] L.D. Faddeev, *Mathematical Aspects of the Three-Body Problem in the Quantum Scattering Theory* (Steklov Mathematical Institute, Leningrad, 1963) [English translation: Israel Program for Scientific Translation, Jerusalem 1965] ; E.W. Schmid, H. Ziegelmann, *The Quantum Mechanical Three-Body Problem* (Pergamon, Oxford, 1974); R.G. Newton, *Scattering Theory of Waves and Particles* (McGraw-Hill, New York, 1966)

[8.12] W. Glöckle, *The Quantum Mechanical Few-Body Problem* (Springer, Berlin, Heidelberg, 1983)

[8.13] G.L. Payne, Lecture Notes in Physics **273** (1987) 64

[8.14] W. Glöckle, Nucl. Phys. **A381** (1982) 343

[8.15] J.L. Friar, B.F. Gibson, D.R. Lehman, and G.L. Payne, Phys. Rev. **C37** (1988) 2859

[8.16] R.B. Wiringa, R.A. Smith, and T.A.Ainsworth, Phys. Rev. **C29** (1984) 1207

[8.17] T. Sasakawa, Few-Body Systems Suppl. **1** (1986) 104; R.A. Brandenburg et al., Phys. Rev. **C37** (1988) 781; Phys. Rev. **C37** (1988) 1245

[8.18] J. Stoer, *Einführung in die numerische Mathematik*, (Springer, Berlin, Heidelberg, 1976)

[8.19] W. Glöckle, G. Hasberg, and A.R. Neghabian, Z. Phys. **A305** (1982) 217

[8.20] R.A. Malfliet and J.A. Tjon, Nucl. Phys. **A127** (1969) 161; J.L. Friar, B.F. Gibson, and G.L. Payne, Z. Phys. **A301** (1981) 309

9. Variational Monte Carlo Techniques in Nuclear Physics

J.A. Carlson and R.B. Wiringa

9.1 Introduction

Variational Monte Carlo techniques have proven to be a very powerful tool for studying quantum systems in the physics community as a whole, and particularly within nuclear physics [9.1–4]. The most important applications in nuclear physics are to few-body systems, either at the nucleon or the quark level. A variety of nuclear properties have been calculated with Monte Carlo methods, including the binding energies, electromagnetic form factors, response functions, and asymptotic properties of the wave functions for three- and four-body nuclei. One can also study the Hamiltonian itself, in particular the effects of three-nucleon interactions in light nuclei. Finally, one can examine low-energy scattering states and electromagnetic and weak transitions with these same methods.

One can also employ variational Monte Carlo methods to study the hadronic spectra in a variety of quark models. The flux-tube model, in particular, has been studied extensively with this method [9.5,6]. In addition to the baryon spectra, one can study the decays of these states, and also the spectra of so-called "exotic" four- and six-quark states. Although we will not cover this subject in any detail here, the computational principles involved are exactly the same as those used to study light nuclei. The major differences are simply in the basis of states employed, and, of course, in the interactions used to model these systems.

Two ingredients are necessary to perform any meaningful variational calculation. The first is good physical insight into the structure of the system. Variational studies are always limited by the form chosen for the wave function, and consequently care must be taken when drawing conclusions based upon its detailed properties. Although it may be easy to estimate the statistical error arising from a Monte Carlo calculation, it is much more difficult to determine the sensitivity to the parametrization.

Given an accurate form for the trial wave function, one must still be able to evaluate the integrals with sufficient accuracy to perform a variational calculation. This problem is far from trivial, especially in the context of nuclear physics, where the interactions are highly state-dependent. A great deal of work has been done on $A = 3$ and 4 nuclei, including studies of the three-nucleon interaction and electromagnetic form factors. In addition, the $A = 5$ problem [9.7] has been studied with simple generalizations of the variational Monte Carlo algorithms presented in this chapter. These methods should allow one to treat nuclei up to $A = 8$, and also to study light hypernuclei with realistic hyperon–nucleon interaction models. Investigations in both of these areas are currently underway.

In addition, variational wave functions of the same form are believed to accurately describe the properties of heavier nuclei. Cluster expansion techniques [9.8] have recently been developed which should allow one to study heavier nuclei such as ^{16}O.

The ground states of light nuclei have also been studied with Green's function Monte Carlo (GFMC) methods, which are very similar to the variational techniques described here. GFMC solves for the ground state of a quantum system by calculating

$$\Psi_0 = \exp\{-H\tau\}\Psi_T, \tag{9.1}$$

where Ψ_T is the variational wave function, Ψ_0 is the exact ground state, and τ is sufficiently large to project out all excited states. Various formulations of this method are described in some detail in Refs. [9.9–12], and its specific application to nuclear physics problems in Refs. [9.13,14].

9.2   General Methods

All variational Monte Carlo calculations rely upon a few simple ideas which we can describe only briefly. The interested reader is invited to consult Refs. [9.15,16] for examples of variational calculations in other areas of physics. Nearly all of these calculations rely upon the Metropolis algorithm [9.17] for generating a set of points with a specific probability distribution. This technique, as well as many other important general Monte Carlo topics, are covered in Ref. [9.18].

To study the ground state of a quantum system, one can make an ansatz for the wave function $\Psi\{\alpha\}$, where $\{\alpha\}$ represents a set of variational parameters that are to be optimized. The expectation value of the Hamiltonian with this wave function gives an estimate of the ground-state energy

$$E\{\alpha\} = \frac{\langle\Psi|H|\Psi\rangle}{\langle\Psi|\Psi\rangle}, \tag{9.2}$$

where $E\{\alpha\}$ is greater than the true ground-state energy E_0. By minimizing $E\{\alpha\}$ with respect to each of the parameters, one obtains an approximation to both E_0 and the ground-state wave function. The error in the estimated ground-state energy is second order in the difference between the trial and true ground-state wave functions, so it is often quite accurate. Expectation values of other observables, however, do not share this property.

It has occasionally proven useful to minimize other quantities, for example the difference

$$\langle H^2 \rangle - \langle H \rangle^2. \tag{9.3}$$

This difference is precisely zero for any true eigenstate of the Hamiltonian, so that in principle it can also be used to determine the excited states, assuming that one has a sufficiently accurate starting trial function. This technique also can be useful for choosing optimum trial wave functions for Green's function Monte Carlo calculations.

Variational Monte Carlo calculations determine $E\{\alpha\}$ by writing it as

$$\langle H \rangle = \frac{\int d\boldsymbol{R} \dfrac{\Psi^\dagger(\boldsymbol{R}) H \Psi(\boldsymbol{R})}{W(\boldsymbol{R})} W(\boldsymbol{R})}{\int d\boldsymbol{R} \dfrac{\Psi^\dagger(\boldsymbol{R}) \Psi(\boldsymbol{R})}{W(\boldsymbol{R})} W(\boldsymbol{R})}. \tag{9.4}$$

$W(\boldsymbol{R})$ is a probability distribution usually taken to be just $\Psi^\dagger(\boldsymbol{R}) \Psi(\boldsymbol{R})$. The value of this expression is that it can be easily evaluated through Metropolis Monte Carlo techniques, which can produce a set of points in the $3A$-dimensional space proportional to $W(\boldsymbol{R})$. Once this set of points $\{\boldsymbol{R}\}$ is obtained, the expectation value of an operator O may be calculated through:

$$\langle O \rangle = \frac{\sum \dfrac{\Psi^\dagger(\boldsymbol{R}) O \Psi(\boldsymbol{R})}{W(\boldsymbol{R})}}{\sum \dfrac{\Psi^\dagger(\boldsymbol{R}) \Psi(\boldsymbol{R})}{W(\boldsymbol{R})}}, \tag{9.5}$$

where the sum runs over all the points in the set $\{\boldsymbol{R}\}$. This method is sufficient, in principle, to determine the expectation value of any operator O. In practice, its effectiveness depends largely upon the structure of the operator. For example, the true wave function is an eigenstate of the Hamiltonian, so each point in the average over the set $\{\boldsymbol{R}\}$ makes an equal contribution. Therefore, in the limit that the variational wave function approaches the exact eigenstate, there is no statistical error associated with Monte Carlo evaluations of $\langle H \rangle$.

In reality, of course, there is always a statistical error, since we do not know the exact wave function. This error can be estimated by making use of the central limit theorem. If there are a sufficiently large number of statistically independent samples in the set $\{\boldsymbol{R}\}$, the distribution of "local" averages obtained in a Monte Carlo evaluation is simply a Gaussian centered on the true value. A "local" average is simply the average over a sufficiently large set of the points in $\{\boldsymbol{R}\}$, and thus the distribution of local averages obtained during a simulation can be used to estimate the statistical error of the calculation:

$$\sigma = \sqrt{\frac{\langle O^2 \rangle - \langle O \rangle^2}{N - 1}}, \tag{9.6}$$

where N is the number of local averages, and σ represents a one-standard-deviation error estimate. Authentic two-standard-deviation effects will occur approximately in one out of twenty calculations.

Great care must be taken to ensure that these local averages are indeed statistically independent and also that any dependence on initial conditions has disappeared. Apparently two-standard-deviation effects appear very often in the literature, often owing to neglect of the terms "sufficiently large" and "statistically independent". These phrases are meaningful only in the context of a specific problem, and their precise values must be determined by experiment in each new calculation. They also depend strongly upon the operator of interest, and not simply on the ground-state wave function. Other error-analysis methods, such as "bootstrapping", are available in cases where the statistical results are more limited. These techniques can be very useful, and they do not rely upon a Gaussian distribution of statistical errors. In general, though, they are unnecessary for analyzing results in light nuclei.

Metropolis Monte Carlo algorithms generate a set of points $\boldsymbol{R}$ with probability distribution $W(\boldsymbol{R})$ through the following basic algorithm:

1. Given a $3A$-dimensional vector $\boldsymbol{R}$, generate a new vector $\boldsymbol{R'}$ with a "transition probability" $T(\boldsymbol{R} \Rightarrow \boldsymbol{R'})$. In the simplest case, T is taken to be a constant within a $3A$-dimensional cube surrounding the point $\boldsymbol{R}$.

2. Calculate the quantities $W(\boldsymbol{R}), W(\boldsymbol{R'}), T(\boldsymbol{R} \Rightarrow \boldsymbol{R'})$, and $T(\boldsymbol{R'} \Rightarrow \boldsymbol{R})$, the transition probability for the reverse step. The acceptance probability is given by the expression

$$P(\boldsymbol{R} \Rightarrow \boldsymbol{R'}) = \min\{1, W(\boldsymbol{R'})T(\boldsymbol{R'} \Rightarrow \boldsymbol{R})/W(\boldsymbol{R})T(\boldsymbol{R} \Rightarrow \boldsymbol{R'})\}. (9.7)$$

3. If the move is accepted, set $\boldsymbol{R} = \boldsymbol{R'}$ and return to step one. Otherwise, discard the point $\boldsymbol{R'}$ and generate the next move from the original $\boldsymbol{R}$.

This algorithm satisfies the condition known as "detailed balance"; that is, that the combined probability for a move from $\boldsymbol{R}$ to $\boldsymbol{R'}$ is equal to the probability for the reverse move. By combined probability one means the product of the probability of being at $\boldsymbol{R}$ to begin with, which we want to be $W(\boldsymbol{R})$, the probability for trying a move from $\boldsymbol{R}$ to $\boldsymbol{R'}$, $T(\boldsymbol{R} \Rightarrow \boldsymbol{R'})$, and the probability for accepting such a move $P(\boldsymbol{R} \Rightarrow \boldsymbol{R'})$. It is trivial to show that the definition of P given in step 2 satisfies this condition.

One must be careful, of course, to use only those points generated after the equilibrium distribution has been reached. This can be determined by tracking various expectation values as a function of the number of Monte Carlo steps. For the simple few-body problems considered here, the equilibrium distribution is usually reached fairly rapidly (a few hundred steps), and for most operators statistically independent samples will be obtained within at most a similar number of steps.

Instead of using a constant (within a prescribed volume V) for the transition probability T, one can use information about the derivatives of W to obtain a more efficient sampling of configuration space. For a given volume V, these "forced bias" [9.18] algorithms give, on average, a higher acceptance probability P. One need only take care that $P(\boldsymbol{R} \Rightarrow \boldsymbol{R}')$ is never less than zero for any $\boldsymbol{R}$ and $\boldsymbol{R}'$. The step size used in the algorithm should be adjusted to minimize the computer time necessary to generate statistically independent results. The lore in this field holds that an acceptance ratio of roughly fifty percent usually gives good results.

Finally, one can often use so-called "expected-value" [9.18] methods to reduce the variance of a calculation. Instead of simply averaging over the set of points $\boldsymbol{R}$ reached in step 1 of the algorithm, one can average at each step P times the value at point $\boldsymbol{R}'$ with $(1 - P)$ times the value at the old point $\boldsymbol{R}$. This, of course, gives the same contribution on average to the integral in question. This technique is primarily useful when the quantity of interest is rather cheap to compute compared to the evaluation of W (here the square of the wave function).

9.3 Interactions and Wave Functions

Realistic treatments of problems in nuclear physics typically involve highly state-dependent interactions. When studying hadron spectra, one is dealing with operators based upon "one-gluon-exchange" interactions [9.19]. We will concentrate on few-nucleon problems, for which the Hamiltonian may be written as

$$H = \sum_i \frac{-\hbar^2}{2m} \nabla_i^2 + \sum_{i<j} \sum_k V^k(r_{ij}) O_{ij}^k + \sum_{i<j<k} V_{ijk}. \tag{9.8}$$

The operators O_{ij}^k in realistic nucleon–nucleon interactions (Argonne v_{14} [9.20], Paris [9.21], Nijmegen [9.22], Reid [9.23], etc.) may be taken to be

$$O_{ij}^k = [1, \sigma_i \cdot \sigma_j, S_{ij}, (L \cdot S)_{ij}, L_{ij}^2, L_{ij}^2 \sigma_i \cdot \sigma_j, \nabla_{ij}^2, \nabla_{ij}^2 \sigma_i \cdot \sigma_j, (L \cdot S)_{ij}^2]$$
$$\otimes [1, \tau_i \cdot \tau_j]. \tag{9.9}$$

These interactions are generally fit to nucleon–nucleon scattering data and deuteron properties, and include the one-pion-exchange interaction at long distances. At intermediate and short distances, they are more phenomenological, often assuming a one-boson-exchange structure.

There are significant differences between the various interaction models; for instance the D-state of the deuteron ranges from 5.2% to 6.5%. We cannot describe the individual models in detail, but we note that the same

methods can be applied to each of the interactions. The three-body interactions are expected to be of some importance in nuclear physics, simply
because the energy scale for internal excitations of the nucleon is fairly small
(hundreds of MeV). Various models for the three-nucleon interactions have
been proposed [9.2,24]; these models incorporate the operator structure of
the two-pion-exchange–three-nucleon interaction at long distances.

The variational wave functions employed to study these Hamiltonians
have the following general structure:

$$\Psi = \left(S \prod_{i<j} F_{ij} \right) \Phi. \tag{9.10}$$

The F_{ij} are pair-correlation operators and Φ is an antisymmetrized product
of one-body states. The F_{ij} are generally noncommuting so the product
must be symmetrized. The combination of strong short-ranged repulsion
and strong state dependence in the interaction makes this a good choice
for the trial function. The product of one-body states Φ incorporates the
correct total angular momentum and isospin for the state of interest. We
can also choose it to be independent of the center of mass in order to insure
the translational invariance of the full wave function. For three- and four-
body nuclei Φ is simply taken to be a product of spin–isospin states with
no spatial dependence:

$$\Phi(\,^3\text{He}) = \mathcal{A}|\uparrow n \ \downarrow p \ \uparrow p\rangle\,; \tag{9.11}$$

$$\Phi(\,^4\text{He}) = \mathcal{A}|\uparrow n \ \downarrow n \ \uparrow p \downarrow p\rangle\,. \tag{9.12}$$

In this case, the correct asymptotic conditions on the wave function are
imposed as conditions on the pair correlations. Of course, other choices are
possible, and the choice of forms for Φ and the pair correlations are closely
related.

The pair correlations are chosen to reflect the influence of the two-body
potential at short distances, while satisfying asymptotic boundary conditions of single-particle separability [9.3,4]. This is achieved by solving six
Schrödinger-like equations in S, T channels: two single-channel equations
for $S = 0$ states, and two coupled-channel equations in the $S = 1$ states.
These equations give four central $f_{S,T}$ and two tensor $f_{t,T}$ functions that can
then be projected into operator form. The six equations are the low-density
limit of those used in nuclear matter studies [9.25] (without spin–orbit correlations):

$$-\frac{\hbar^2}{m}\left[(f_{\text{S,T}}\,r^{l+1})'' - \frac{l(l+1)}{r^2}(f_{\text{S,T}}\,r^{l+1})\right] \tag{9.13}$$
$$+\ (\bar{v}_{\text{S,T}}\{\alpha\} + \lambda_{\text{S,T}})(f_{\text{S,T}}\,r^{l+1}) + 8(\bar{v}_{\text{t,T}}\{\alpha\} + \lambda_{\text{t,T}})(f_{\text{t,T}}\,r^{l+1})\delta_{S1} = 0,$$

and

$$-\frac{\hbar^2}{m}\left[(f_{\text{t,T}}\,r^{l+1})'' - \frac{6+l(l+1)}{r^2}(f_{\text{t,T}}\,r^{l+1})\right]$$
$$+\ (\bar{v}_{\text{S,T}}\{\alpha\} + \lambda_{\text{S,T}} - 2(\bar{v}_{\text{t,T}}\{\alpha\} + \lambda_{\text{t,T}}) - 3\bar{v}_{\text{b,T}}\{\alpha\})(f_{\text{t,T}}\,r^{l+1})$$
$$+\ (\bar{v}_{\text{t,T}}\{\alpha\} + \lambda_{\text{t,T}})(f_{\text{S,T}}\,r^{l+1}) = 0, \tag{9.14}$$

where $''$ denotes a second derivative, the $\bar{v}\{\alpha\}$ are quenched potentials characterized by a variational parameter α (including the spin–orbit potentials $\bar{v}_{\text{b,T}}\{\alpha\}$), and $l = 0(1)$ for $S,T = 0,1$ and $1,0$ ($0,0$ and $1,1$). The boundary conditions for $f_{\text{S,T}}$ and $f_{\text{t,T}}$ are

$$f_{\text{S,T}}(r \to 0) = \text{constant}$$
$$f_{\text{t,T}}(r \to 0) = 0 \tag{9.15}$$
$$f_{\text{S,T}}(r \to \infty) = \left[\frac{\exp(-k_{S,T}\,r)}{r}\right]^{1/(A-1)}$$
$$f_{\text{t,T}}(r \to \infty) = \eta_T\left(1 + \frac{3}{k_{S,T}\,r} + \frac{3}{(k_{S,T}\,r)^2}\right)\left[\frac{\exp(-k_{S,T}\,r)}{r}\right]^{1/(A-1)},$$

with

$$k_{S,T} = \left[\frac{A-1}{A}\frac{2m}{\hbar^2}E_{S,T}\right]^{1/2}. \tag{9.16}$$

The four separation energies $E_{S,T}$ and two tensor/central ratios η_T are variational parameters. The Lagrange multipliers λ_x are radial functions consisting of two parts: the long-range part $\Lambda_x(r)$ is fixed by the asymptotic behavior of f_x and is cut off at short distances by an exponential function, while the short-range part is a Woods–Saxon function multiplied by a constant Γ_x;

$$\lambda_x = \Gamma_x\left[1 + \exp\left(\frac{r-R_x}{a_x}\right)\right]^{-1} + \Lambda_x(r)\{1 - \exp[-(r/c_x)^2]\}. \tag{9.17}$$

These constants are determined by solving the differential equations subject to the boundary conditions given above. The Woods–Saxon and cut-off constants (R_x, a_x, and c_x) are additional variational parameters. After solving the differential equations, the correlations are recast into operator form:

$$\sum_x f_x(r_{ij}) P_{ij}^x = \sum_{k=1}^{6} f^k(r_{ij}) O_{ij}^k = f^c(r_{ij}) \left[1 + \sum_{k=2}^{6} u^k(r_{ij}) O_{ij}^k \right], \qquad (9.18)$$

where the P^x are channel projection operators and the operators O_{ij}^k are

$$O_{ij}^k = 1, \tau_i \cdot \tau_j, \sigma_i \cdot \sigma_j, \sigma_i \cdot \sigma_j \tau_i \cdot \tau_j, S_{ij}, S_{ij} \tau_i \cdot \tau_j. \qquad (9.19)$$

Empirically it has been necessary to introduce three-body correlations which modify the strength of the two-body spin-dependent correlations whenever more than two particles are near each other. We have introduced a set of three-body correlations

$$f_3 = \prod_{k \neq i,j} \left[1 - t_1 \left(\frac{r_{ij}}{R} \right)^{t_2} \exp(-t_3 R) \right], \qquad (9.20)$$

with $R = r_{ij} + r_{ik} + r_{jk}$. These correlations multiply the u^k in the equation above. The parameters t_1, t_2, and t_3 are determined variationally, and may in principle be different for each of the five correlation functions. In practice, we keep only two, one for the tensor operators and another for the nontensor terms.

9.4   Monte Carlo

Once we have defined a wave function and an interaction, we need to develop a code which will evaluate the expectation value of a strongly state-dependent Hamiltonian. This is fairly simple in principle, as we can rewrite (9.4) as

$$\langle H \rangle = \frac{\int d\boldsymbol{R} \frac{\sum_{kl} \Psi_k^\dagger (\boldsymbol{R}) H_{kl} \Psi_l (\boldsymbol{R})}{W(\boldsymbol{R})} W(\boldsymbol{R})}{\int d\boldsymbol{R} \sum_{kl} \frac{\Psi_k^\dagger (\boldsymbol{R}) \Psi_l (\boldsymbol{R})}{W(\boldsymbol{R})} W(\boldsymbol{R})}, \qquad (9.21)$$

where the sums over k and l run over all spin–isospin states. In principle, one could define a weight which is a function of k and l, and use Monte Carlo to evaluate these sums as well as the integrals in configuration space. However, this method will produce large statistical errors, because one is no longer making use of the cancellations that arise because the wave function is an eigenstate of the Hamiltonian. By performing the sums over spin and isospin states explicitly at each point in the walk rather than using Monte Carlo, the variance of the calculation can be greatly reduced. This explicit summation, though, limits the applicability of the algorithm to light ($A \leq$ 8) nuclei.

The form of variational wave function (9.10) implies the possibility of a further gain in efficiency. The symmetrization operator S indicates a sum over all orders of the pair correlation operators. For example,

$$S(F_{12}F_{13}F_{23}) = \frac{1}{\sqrt{6}}[F_{12}F_{13}F_{23} + F_{13}F_{12}F_{23} + F_{23}F_{12}F_{13} + ...]. \qquad (9.22)$$

Monte Carlo techniques can be used to sample the terms in the sum, choosing the order of operators independently for $\Psi(\boldsymbol{R})$ and $\Psi^{\dagger}(\boldsymbol{R})$:

$$
\begin{aligned}
\Psi^{\dagger}(\boldsymbol{R}) &= \sum_{p} \Psi_{p}^{\dagger}(\boldsymbol{R}), \\
\Psi(\boldsymbol{R}) &= \sum_{q} \Psi_{q}(\boldsymbol{R}).
\end{aligned}
\qquad (9.23)
$$

In addition, the weight function W can be made a function of p and q. We have chosen

$$W_{pq}(\boldsymbol{R}) = |\mathrm{Re}\langle \Psi_{p}^{\dagger}(\boldsymbol{R})|\Psi_{q}(\boldsymbol{R})\rangle|, \qquad (9.24)$$

where the brackets indicate a sum over all spin and isospin states. The absolute magnitude is required in order to be certain that the weight function is never less than zero. In light nuclei, the real part of the product $\langle \Psi_{p}^{\dagger}\Psi_{q}\rangle$ is positive for any p and q for reasonable choices of correlation functions.

The statistical error per sample in configuration space is increased by sampling the order of operators. However, the time saved by evaluating only one term in the sum increases the efficiency of the calculation, especially for $A > 3$. The complete expression for the expectation value of an operator O (analogous to (9.5)) is

$$\langle O \rangle = \frac{\dfrac{\langle \Psi_{p}^{\dagger}(\boldsymbol{R})\, O\Psi_{q}(\boldsymbol{R})\rangle}{W_{pq}(\boldsymbol{R})}}{\dfrac{\langle \Psi_{p}^{\dagger}(\boldsymbol{R})\, \Psi_{q}(\boldsymbol{R})\rangle}{W_{pq}(\boldsymbol{R})}}. \qquad (9.25)$$

$W_{pq}(\boldsymbol{R})$ is the weight function used to generate the coordinates in configuration and permutation space, and the brackets indicate sums over all spin–isospin states.

It is often desirable to use a given set of configurations, obtained with one W_{pq}, to evaluate the energy of several different variational wave functions. Although this procedure produces a biased estimate of the energy (since both the numerator and denominator will contain a statistical error), it can be very valuable in minimizing the energy. There are often large cancellations in the difference in energy of two wave functions, and these cancellations can be exploited by using this "reweighting" technique. This method is most valuable for small changes in the wave function, in which case the bias is small. Once the variational parameters are optimized, one should always do a long unbiased run to produce the best estimate of the energy. In this manner one also eliminates another common problem with variational Monte Carlo calculations: an artificial lowering of the energy estimate associated with using the same run to optimize the parameters and evaluate the energy.

A convenient basis of spin–isospin states is provided by those sets in which each particle has a definite third component of spin and isospin. The Hamiltonian and pair-correlation operators are sparse matrices in this basis, since two-body correlations or interactions can only change the spins or isospins of two nucleons at a time.

For example, the operator $\sigma_i \cdot \sigma_j$ may be written as

$$\sigma_i \cdot \sigma_j = 2P_{ij}^\sigma - 1, \tag{9.26}$$

and the tensor operator

$$S_{ij} = 3(\sigma_i^+ \hat{r}^- + \sigma_i^- \hat{r}^+ + \sigma_i^0 \hat{r}^0)(\sigma_j^+ \hat{r}^- + \sigma_j^- \hat{r}^+ + \sigma_j^0 \hat{r}^0) - 2P_{ij}^\sigma + 1, \tag{9.27}$$

where

$$\begin{array}{llll}
\sigma^+ &= (\sigma^x + i\sigma^y)/2 & \sigma^- &= (\sigma^x - i\sigma^y)/2 \quad \sigma^0 = (\sigma^z), \\
\hat{r}^+ &= \hat{r}^x + i\hat{r}^y & \hat{r}^- &= \hat{r}^x - i\hat{r}^y \quad\;\; \hat{r}^0 = \hat{r}^z,
\end{array} \tag{9.28}$$

P_{ij}^σ interchanges spins i and j, and σ^+ and σ^- convert down spins to up and up to down, respectively.

Within the computer, it is convenient to represent the state Ψ_p as a two-dimensional complex array containing the coefficients associated with each spin–isospin state. The indices of the array represent, in a simple binary fashion, the spin and isospin states of the system. For example, the spin indices are given by

$$\begin{array}{rcl}
\downarrow\downarrow\downarrow\downarrow &=& (0000)_{\text{base 2}} + 1 = 1 \\
\downarrow\downarrow\downarrow\uparrow &=& (0001)_{\text{base 2}} + 1 = 2 \\
\downarrow\downarrow\uparrow\downarrow &=& (0010)_{\text{base 2}} + 1 = 3 \\
\downarrow\downarrow\uparrow\uparrow &=& (0011)_{\text{base 2}} + 1 = 4 \\
\uparrow\uparrow\uparrow\uparrow &=& (1111)_{\text{base 2}} + 1 = 16.
\end{array} \tag{9.29}$$

At this point, it is simple to create a table containing the results of acting with a spin-exchange operator for any pair acting on any initial state. Similarly, one can construct tables for the spin-flip $(\sigma_i^+ + \sigma_i^-)$ operator.

The isospin states are handled similarly, but not all bit patterns are retained because of charge conservation. Thus, there are only 6 isospin states in an alpha particle, but 16 spin states. One could further reduce the storage requirements by explicitly invoking isospin conservation, but this would necessitate a more complicated matrix structure for the isospin-exchange operator.

Once the operator tables are constructed, it is easy to compute the result of any operator of the form

$$O = \sum_{k=1}^{6} c_k O_{ij}^k \qquad (9.30)$$

acting on an arbitrary initial state. In this expression, c_k are complex coefficients and O^k are the six operators given in (9.19). One can easily compute the wave function Ψ_q by repeated operations of this sort, applying each pair operator consecutively. The expectation values of the static potential terms can be evaluated similarly.

The expectation value of the kinetic energy as well as other momentum-dependent terms in the interaction must also be evaluated. The kinetic energy and $L \cdot S$ terms require both the first derivatives of the wave function and the diagonal second derivatives to be computed. These are obtained simply by moving each particle a small distance ϵ in both the positive and negative directions along each axis:

$$\nabla_j^i \Psi\{R\} = [\Psi\{R + \epsilon \hat{r}_j^i\} - \Psi\{R - \epsilon \hat{r}_j^i\}]/[2\epsilon]$$
$$\nabla_j^{i^2} \Psi\{R\} = [\Psi\{R + \epsilon \hat{r}_j^i\} + \Psi\{R - \epsilon \hat{r}_j^i\}] - 2\Psi\{R\}]/[\epsilon^2]. \qquad (9.31)$$

In these expressions, i represents a direction ($x, y,$ or z), and j represents the particle. The $L \cdot S$ operator can be rewritten in terms of σ^+, σ^-, and σ^0, as was done above for the tensor operator. In the interests of brevity, the program included with this chapter does not treat L^2 or $(L \cdot S)^2$ operators, and so is limited to simpler interaction models such as the Reid v_8, which is a simplification of the full Reid interaction [9.23]. The operators which contain two derivatives may be included either by using integration by parts to apply one derivative operator to $\Psi(R)$ and one to $\Psi^\dagger(R)$, or by direct numerical calculation of the remaining second partial derivatives. This latter technique is necessary when trying to compute energy-weighted sum rules [9.26].

9.5   Results

The optimized variational parameters we have obtained for the triton with the Reid v_8 interaction are listed in Table 9.1. These parameters were determined after a series of runs, many of which employed the reweighting techniques described above. Typically, twenty to thirty runs of several thousand configurations are necessary to optimize the parameters for each new interaction model. With reweighting techniques, it is fairly easy to evaluate the difference in energy between two similar wave functions to within a few hundredths of an MeV.

Once the optimum wave function has been determined, a set of Monte Carlo calculations should be undertaken to determine all of the expectation values. For the three-body problem, ten to twenty thousand configurations seem to provide reasonable statistical accuracy for the energy and one-body densities. Ten thousand configurations take roughly 30 minutes of CPU

Table 9.1. Triton Variational Parameters

	$^1S_0$	$^1P_1$	$^3S_1$	^3P_J	$^3D_1$	^3F_J
$E_{S,T}$	6.0	2.0	12.0	6.0		
η_T					.026	$-.010$
c_x	1.0	1.0	3.0	3.0	2.0	2.0
a_x	0.4	0.4	0.4	0.4	0.4	0.4
R_x	1.0	1.0	2.8	2.8	3.6	3.6
α_x	1.00	0.92	0.92	0.92	0.92	0.92
	$t1_{S,T}$	$t1_{t,T}$	$t2_{S,T}$	$t2_{t,T}$	$t3_{S,T}$	$t3_{t,T}$
	10.0	10.0	4.0	4.0	0.05	0.05

time on a one megaflop computer. Of course, this calculation can be split up into many small runs, and the energies and statistical error determined from these results.

The results obtained with this wave function are summarized in Table 9.2 and Fig. 9.1. The table presents all of the energy expectation values, as well as point-particle rms radii of the neutron and proton density. In this table, V_{ij} gives the total nucleon–nucleon potential energy, and T_i is the total kinetic energy. There is a strong cancellation in the total energy owing to the strong repulsive core in the nucleon–nucleon interaction. The two-body potential is also split up into the V_6 contribution and the remaining $L \cdot S$ terms (V_b).

Table 9.2. Triton Results

	Expectation Value	Statistical Error
$T_i + V_{ij}$	-7.30	0.04
$T_i + V_{ij} + V_{ijk}$	-8.44	0.06
T_i	52.4	0.6
V_{ij}	-59.7	0.6
V_6	-60.8	0.6
V_b	1.1	0.1
V_{ijk}	-1.14	0.03
$V3_a$	-1.04	0.02
$V3_c$	-0.68	0.02
$V3_u$	0.58	0.02
$\langle r_i^2 \rangle$ proton	1.620	0.004
$\langle r_i^2 \rangle$ neutron	1.766	0.004

In addition, the contribution of the various pieces of the Urbana model VII three-nucleon-interaction are given in Table 9.2. The commutator and

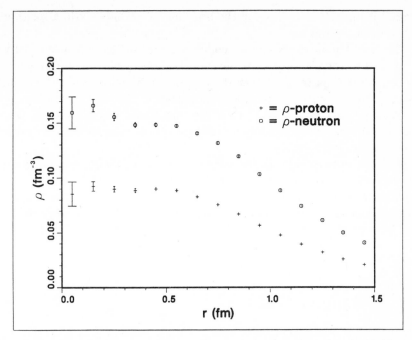

Fig. 9.1. Point proton and neutron densities in the triton for the Reid v_8 + Urbana VII TNI interaction.

anticommutator pieces of the two-pion-exchange TNI are listed as $V3_c$ and $V3_a$ respectively, and the short-ranged repulsive piece is $V3_u$. Each of these terms is a very small fraction of the total potential energy, yet they constitute a significant part of the binding energy of the three- and four-nucleon systems. The neutron and proton one-body densities are given in Fig. 9.1. The neutron and proton form factors can easily be folded in to get the charge density and then Fourier transformed to get the impulse approximation to the charge form factor. Meson exchange currents play a very important role in getting the correct charge form factor, even at relatively moderate values of the momentum transfer. The same Monte Carlo methods can be used to determine these expectation values. Schiavilla et al. have obtained excellent agreement with the charge form factors employing these methods [9.27].

Many other physically interesting quantities can be calculated, including non-energy-weighted [9.28] and energy-weighted Coulomb sum rules [9.26], momentum distributions, etc. The time required for any calculation depends not only upon the desired accuracy and the available computer capacity, but also on the operator one is interested in. The expectation values of nonlocal operators, for example the momentum distribution of nucleons in the nucleus, are particularly prone to high statistical error. In such cases it is important to develop the best possible Monte Carlo algorithm to minimize the variance.

Taking a simple example, this code calculates the one-particle density as a function of the distance from the center of mass. The statistical errors in the density are quite large near the origin simply because of the r^2 phase-space factor. If one is particularly interested in the density at this point, it would be very profitable to redo the calculation with a different weight function W, one which emphasizes configurations where at least one particle is near the center of mass.

Current work in light nuclei centers on the possibilities of treating larger systems and the dynamical properties of light nuclei. These methods hold promise in many avenues of nuclear physics. In particular, one can study parts of nuclear physics related to astrophysics, including neutron-rich nuclei and low-energy reactions. In addition, successful calculations of dynamical properties are extremely important to our understanding of current and future electron-scattering experiments.

9.6   Computer Code

The program presented here is a somewhat simplified version of our standard few-nucleon code. It contains one two-nucleon potential (Reid v_8) and one three-nucleon potential (Urbana VII) and solves for the ground-state binding energy and density distribution for either three- or four-body nuclei. It can also save random walks generated with one trial function and then compute energy differences with a different trial function. It is written in ANSI standard FORTRAN 77 (with the exception of using lowercase letters) and we have verified that it runs on a number of different computers. The code includes performance monitoring, and benchmark information is presented in Table 9.3.

In the following, we describe the general structure of the code and point out some of the specific features. Details of the specific variables and actions of the program are given as comments in the code listing.

The computer code includes a main routine, called NUCLEI, and some 22 subroutines and functions. The code requires double-precision arithmetic and includes complex variables. Variables which specify the choice between the three- or four-body problems are set in PARAMETER statements that are shared by most of the program elements. (This improves efficiency by specifying many DO-loop lengths at compilation time.) A small data file is read in from unit NIN and output is sent to unit NOUT.

Initialization includes calls to a function TIMER, which starts timing data for monitoring performance, and a subroutine HEADER, which identifies the system, compiler, time, and date. These calls are machine specific; here we provide code (which is selected by removing appropriate comment prefixes in the subroutines) for the machines in Table 9.3. Data is read (and

Table 9.3. Code Performance and Timing. "Time" is the number of seconds required to compute one energy sample, including moving for a new walk. "Speed" is in MFLOPS.

Computer	Operating system / Fortran compiler	^{3}H old walk $T_i + V_{ij}$		^{4}He new walk $T_i + V_{ij} + V_{ijk}$	
		time	speed	time	speed
VAX 11/780	VMS / vax fortran v5.3	0.68	0.15	11.	0.15
Sun-3/160	UNIX / f77 -O -ffpa	0.48	0.21	8.7	0.20
VAX 3100	VMS / vax fortran v5.2	0.25	0.41	4.0	0.43
Sun-4/280	SunOS / f77 -O	0.13	0.80	2.0	0.88
VAX 8700	VMS / vax fortran v5.2	0.10	1.0	1.7	1.0
Sun Sparc 1	SunOS / f77 -O	0.10	1.0	1.7	1.1
IBM 3033	MVS / vs fortran 2.3.0	0.064	1.6	0.96	1.8
IBM 3090/VF	VM / vs fortran 2.3.0	0.030	3.3	0.34	5.1
Cray-2S	CTSS / cft77 3.1	0.0054	19.	0.050	37.
Cray X-MP	UNICOS / cft77 3.1.1.2	0.0037	28.	0.034	51.

printed) specifying the Hamiltonian. Potential functions are set by calls to subroutine POT.

Next, components of the wave function are initialized. The initial un-correlated wave function Φ of (9.11,12) and the spin–isospin matrices which give the result of spin-exchange and spin-flip operations on the wave function (9.26–28) are produced by subroutine SETSPN and its function EXCH. Data specifying the two-body correlation operator is read, and the correlation functions f^c and u^k are generated by the subroutine F6COR according to (9.14–18). An optional printout of the spin-isopin matrices and correlation functions is provided by subroutine WFPRT. Variables for the three-body correlation f_3 of (9.20) are also read.

Variables controlling the random walk are read next. The pseudo-random-number generator is initialized with the call SETRND. The option of saving a random walk by writing a file on unit NRW, or of following a previously saved random walk by reading an existing file, is available. Additional variables are initialized, including the local and global energy sums. A starting wave function is then generated by following the general Metropolis moving procedure (provided by subroutine MOVEEM as discussed below) for a long enough time to randomize the particle positions and order of operators in the correlation-operator product. Alternatively, if a previous walk is being followed, subroutine REMOVE is used instead.

The calculation proceeds by obtaining a position $\boldsymbol{R}$ drawn from the weight function $W_{pq}(\boldsymbol{R})$ of (9.24). This is accomplished by a call to subroutine MOVEEM, which uses the Metropolis algorithm of (9.7). MOVEEM first uses the pseudo-random-number generator RNDNMB to select a new R' and order of operators p' and q'. The use of our own pseudo-random-number generator ensures reproducibility of results when the code is moved from one machine to another. Pair distances and unit vectors are evaluated, and the corresponding functional values f^c and u^k are interpolated by subroutine INTERP. If any pair of particles is too close, or too far apart, for the three-point interpolation to be valid, the move is rejected and a warning counter is incremented. The $\Psi_{p'}^\dagger(R')$ and $\Psi_{q'}(R')$ are then constructed by calls to subroutine WAV. $W_{p'q'}(R')$ is calculated using the function RCDOT, which evaluates the real part of the dot product of the Hermitian conjugate of one vector with a second vector. The weight is compared with the previous $W_{pq}(R)$ and the move is accepted or rejected accordingly. If it is accepted, interpolation variables generated in INTERP are saved with the call SAVEIT. This procedure is repeated a number of times to obtain a statistically independent position. A final call to entry RESTOR makes sure the interpolation data for the last accepted move is available before returning to the main program.

The bulk of computation is in the construction of the wave function performed by WAV, given the particle positions and order of operators. WAV starts with the uncorrelated spin–isospin state Φ and multiplies in the product over pairs of central correlations f^c. Then the product of correlation operators is built up by successive calls to subroutine F6OP in the order specified. F6OP takes an input wave function vector Ψ and functions c_k and returns the result $O\Psi$ where O is the operator of (9.19). The subroutine uses the mathematics of (9.26–28) and is written in real arithmetic for speed. More than 50% of the entire code's floating-point operations are executed inside F6OP. The c_k passed by WAV to F6OP include both the non-central two-body correlations u^k and the central three-body correlation f_3 of (9.20).

Once a new sample $W_{pq}(R)$ has been found, the contribution to any desired expectation value can be calculated. The energy is calculated by subroutines EKIN, VPOT, and TBPOT, which evaluate the kinetic, two-body potential, and three-body potential, respectively. Contributions to the density distribution are incremented with the subroutine RHOR. EKIN computes derivatives of $\Psi(R)$ by differences, as in (9.31), by repeated calls to INTERP and WAV (and F6OP). After a call to subroutine INTRPV to obtain interpolated potential functions, VPOT evaluates the first six operator terms in the potential by another call to F6OP. Spin–orbit terms in the potential are evaluated with the aid of gradients generated and stored by EKIN. Finally the Coulomb potential is evaluated with the charge operator $(1 + \tau_{zi})(1 + \tau_{zj})e^2/r_{ij}$. (A charge form factor is included in our routine.) TBPOT loops over all triples to evaluate the three-body potential contributions. It makes use of two subroutines, STONE and STAC, which have logic very similar to F6OP, but are specialized for the pair operators in the three-nucleon interaction. The three-body force may be written in a form involving the commutators and anticommutators of these pair terms, which are evaluated in the routine STAC.

A running sum is kept of the various components of the energy and of their squares for the calculation of the variance. If an energy difference with a different wave function is being computed, terms for the covariance are also kept. After a certain number of energies are evaluated, a local average and variance (and difference with its variance if appropriate) are computed and printed out, along with a running global average and variance. When the desired total number of configurations have been evaluated, the counters for the density distribution are processed, and the rms radius for protons and neutrons is calculated, and these items are also printed out. The program ends with a printout of the total execution time, number of floating-point operations, which have been tracked with counters in the subroutines, and resulting speed for the code.

The code can be generalized to handle other Hamiltonians, and the wave function can be used to compute other expectation values. The code has been written for execution on a vector processor, but many of the tricks used are also valuable on scalar machines. For nested DO loops, the longest possible ones are placed innermost, and where possible, run over multiple indices. For example, loops over both spin and isospin indices frequently run over $IJ=1,NS*NT$ with functions dimensioned $F(NS,NT)$ indexed as $F(IJ,1)$. (This has the disadvantage of giving error messages if a bounds checker is used.) A number of further improvements can be made that will increase performance, but they have been omitted here for simplicity. Real arithmetic is generally faster than complex, and explicitly real versions of the subroutines STONE and STAC are useful. The kinetic-energy calculation can be sped up significantly on a vector processor by making versions of subroutines WAV and F60P that compute the $6A$ wave functions needed for gradients at the same time, i.e., with the innermost DO loop running over an index $L=1,6*NPART$. Memory-bank conflicts can be reduced on some machines by adding one unit to the size of the leading dimension in many arrays, e.g., dimensioning objects like $F(NS+1,NT)$. These are the primary differences between the code presented here and our regular production code.

9.7   Technical Note

The code is not intended to be run on a microcomputer. Thus we have not tried to convert it to strict ANSI-standard FORTRAN 77. For example, the program uses complex*16 arithmetic which is not available in ANSI-standard FORTRAN 77 and will not be supported by a personal computer.

The diskette contains example input (VARMCH3.INP) and output (VARMCH3.OUT) files which both correspond to studies of the ^3H nucleus with the present program.

9.8   Acknowledgment

J. Carlson would like to thank the U. S. Department of Energy for its support of this work, and R. B. Wiringa would also like to thank the Nuclear Physics Division of the Department of Energy, contract W-31-109-ENG-38.

References

[9.1] J. Lomnitz Adler, V. R. Pandharipande, and R. A. Smith, Nucl. Phys. **A315** (1981) 399

[9.2] J. Carlson, V. R. Pandharipande, and R. B. Wiringa, Nucl. Phys. **A401** (1983) 59

[9.3] R. Schiavilla, V. R. Pandharipande, and R. B. Wiringa, Nucl. Phys. **A449** (1986) 219

[9.4] R. B. Wiringa, to be published.

[9.5] J. Carlson, J. Kogut, and V. R. Pandharipande, Phys. Rev. **D27** (1983) 233

[9.6] J. Carlson, J. Kogut, and V. R. Pandharipande, Phys. Rev. **D28** (1983) 2807

[9.7] J. Carlson, K. E. Schmidt, and M. H. Kalos, Phys. Rev. **C36** (1987) 27

[9.8] S. Pieper, R. B. Wiringa, and V. R. Pandharipande, Phys. Rev. Lett. **64** (1990) 364

[9.9] M. H. Kalos, Phys. Rev. **128** (1962) 1791

[9.10] M. H. Kalos, D. Levesque, and L. Verlet, Phys. Rev. **A9** (1974) 2178

[9.11] K. E. Schmidt, in *Models and Methods in Few-Body Physics*, Lecture Notes in Physics 273 (Springer, Berlin, Heidelberg, 1987)

[9.12] P. J. Reynolds, D. M. Ceperley, B. J. Alder, and W. A. Lester, J. Chem. Phys. **77** (1982) 5593; D. M. Ceperley and B. J. Alder, J. Chem. Phys. **81** (1984) 5833

[9.13] J. Carlson, Phys. Rev. **C38** (1988) 1879

[9.14] J. Carlson, Phys. Rev. **C36** (1987) 2026

[9.15] K. E. Schmidt and J. W. Moskowitz, J. Chem. Phys. **76** (1982) 1064

[9.16] K. E. Schmidt, M. H. Kalos, M. A. Lee, and G. V. Chester, Phys. Rev. Lett **45** (1980) 573

[9.17] N. Metropolis, A. W. Rosenbluth, M. N. Rosenbluth, A. H. Teller, and E. Teller, J. Chem. Phys. **21** (1953) 1087

[9.18] Malvin H. Kalos and Paula A. Whitlock, *Monte Carlo Methods,* Vol. I (Wiley, New York, 1986)

[9.19] A. De Rújula, Howard Georgi, and S. L. Glashow, Phys. Rev. **D12** (1975) 147

[9.20] R. B. Wiringa, R. A. Smith, and T. L. Ainsworth, Phys. Rev. **C29** (1984) 1207

[9.21] M. Lacombe, B. Loiseau, J. M. Richard, R. Vinh Mau, J. Côté, P. Pirés, and R. de Tourreil, Phys. Rev. **C21** (1986) 861

[9.22] M. M. Nagles, T. A. Rijken, and J. J. de Swart, Phys. Rev. **D17** (1978) 768

[9.23] R. V. Reid, Ann. Phys. (N.Y.) **50** (1968) 411

[9.24] S. A. Coon, M. D. Scadron, P. C. McNamee, B. R. Barrett, D. W. E. Blatt, and B. H. J. McKellar, Nucl. Phys. **A317** (1979) 242

[9.25] I. E. Lagaris and V. R. Pandharipande, Nucl. Phys. **A359** (1981) 349

[9.26] R. Schiavilla, A. Fabrocini, and V. R. Pandharipande, Nucl. Phys. **A473** (1987) 290

[9.27] R. Schiavilla, V. R. Pandharipande, and D. O. Riska, Phys. Rev. **C41** (1990) 309

[9.28] R. Schiavilla, et al., Nucl. Phys. **A473** (1987) 267

10. Electron-Scattering Form Factors and Nuclear Transition Densities

H.P. Blok and J.H. Heisenberg

10.1 Introduction

Electron scattering is a very useful method for investigating nuclear structure. Since the electromagnetic interaction is fully known, the only unknowns in the description of the reaction are the nuclear (transition) charge and current densities, which can be determined by analyzing the experimental data. These densities can then be compared to those calculated by some nuclear model in order to test the model wave functions. As model wave functions are often used in some other part of nuclear physics (for instance, to describe proton or pion scattering) it is of importance to know if they are at least consistent with electron-scattering results, if available. This can be done by calculating the transition densities from the wave functions and comparing them to those extracted from the (e,e') results, if the latter are available. Otherwise the densities can be used to calculate the (e,e') cross sections (or form factors) and then these can be compared to the measured data.

The programs described in this chapter calculate transition densities starting from a microscopic description; i.e., in terms of protons and neutrons in the nucleus. Form factors and cross sections are also calculated. The latter is done in the plane wave Born approximation (PWBA), which means that the distortion of the incoming and outgoing electron waves in the Coulomb field of the nucleus is neglected.

The chapter is organized as follows. Section 10.2 explains the formalism used and gives the formulae as used in the programs. A description of the various routines is given in Sect. 10.3. Section 10.4 explains the input to the programs, while the output is discussed in Sect. 10.5.

Details about the formalism and its use for nuclear-structure investigations are given in some review papers [10.1–3].

10.2 Formalism

In electron-scattering experiments an electron with energy E_i is scattered from a target nucleus and the scattered electron of energy E_f is detected at an angle θ. Excitation of the nucleus is manifest in the energy loss $\hbar\omega = E_i - E_f$, so a measurement of the energy spectrum of the scattered electrons determines the excitation spectrum of the nucleus in question. By varying E_i and/or θ, one can measure the cross sections for individual excited states as a function of the momentum transfer q, where $q = k_i - k_f$. This is the key to the determination of the densities, as the cross section at a certain q is basically the Fourier transform of the transition charge and current densities for that state. This means that, if one has data for a sufficient range of q values, one can reconstruct these densities.

In the plane wave Born approximation, the cross section can be written in the form

$$\frac{d\sigma}{d\Omega} = \sigma_p \eta \tag{10.1}$$

$$\times \left[\sum_{\lambda=0}^{\infty} \frac{q_\mu^4}{q^4} |F_\lambda^C(q)|^2 + \left(\frac{q_\mu^2}{2q^2} + \tan^2\frac{\theta}{2} \right) \sum_{\lambda=1}^{\infty} \{|F_\lambda^E(q)|^2 + |F_\lambda^M(q)|^2\} \right].$$

Here the Mott cross section for unit charge σ_p is

$$\sigma_p = \frac{\alpha^2(\hbar c)^2 \cos^2\frac{\theta}{2}}{4E_i^2 \sin^4\frac{\theta}{2}}, \tag{10.2}$$

and the recoil factor η is

$$\eta = \left(1 + \frac{2E_i \sin^2\frac{\theta}{2}}{Mc^2} \right)^{-1}, \tag{10.3}$$

with M being the mass of the target nucleus. [In the programs this (usually small) recoil factor is neglected.] Further, $q_\mu^2 = q^2 - \omega^2$. In these expressions one-photon exchange has been assumed and the rest mass of the electron is neglected.

The nuclear structure enters the cross section through the longitudinal form factor F^C and the transverse form factors F^E and F^M. These form factors are functions of the momentum transfer q only. They can be expressed as Fourier–Bessel transforms of the nuclear charge and current densities:

$$F_\lambda^C(q) = \sqrt{4\pi} \sqrt{\frac{2J_f+1}{2J_i+1}} \int_0^\infty \rho_\lambda(r) j_\lambda(qr) r^2 dr, \tag{10.4}$$

$$F_\lambda^E(q) = \sqrt{4\pi} \sqrt{\frac{2J_f+1}{2J_i+1}} \int_0^\infty \left[\sqrt{\frac{\lambda+1}{2\lambda+1}} J_{\lambda\lambda-1}(r) j_{\lambda-1}(qr) \right.$$
$$\left. - \sqrt{\frac{\lambda}{2\lambda+1}} J_{\lambda\lambda+1}(r) j_{\lambda+1}(qr) \right] r^2 dr, \tag{10.5}$$

and

$$F_\lambda^M(q) = \sqrt{4\pi} \sqrt{\frac{2J_f+1}{2J_i+1}} \int_0^\infty J_{\lambda\lambda}(r) j_\lambda(qr) r^2 dr. \tag{10.6}$$

The transition densities are the reduced matrix elements of the charge or current operator between the initial and final nuclear states:

$$\rho_\lambda(r) = \int \langle \psi_f \| \rho_{\mathrm{op}}(\boldsymbol{r}) Y_\lambda(\hat{\boldsymbol{r}}) \| \psi_i \rangle \, \mathrm{d}\hat{r} \; ; \tag{10.7}$$

$$J_{\lambda\lambda'}(r) = \frac{\mathrm{i}}{c} \int \langle \psi_f \| \boldsymbol{J}_{\mathrm{op}}(\boldsymbol{r}) \cdot \boldsymbol{Y}_{\lambda\lambda'1}(\hat{\boldsymbol{r}}) \| \psi_i \rangle \, \mathrm{d}\hat{r}, \tag{10.8}$$

where the Y_λ are the usual spherical harmonics and $\boldsymbol{Y}_{\lambda\lambda'1}$ are the vector spherical harmonics. The Wigner–Eckart theorem is used in the form

$$\langle J_f M_f | T_{\lambda\mu} | J_i M_i \rangle = (J_i M_i \lambda\mu | J_f M_f) \langle J_f \| T_\lambda \| J_i \rangle. \tag{10.9}$$

The well-known reduced electromagnetic transition probabilities are given in terms of these densities as

$$B(E\lambda) = \frac{2J_f + 1}{2J_i + 1} \left[\int_0^\infty \rho_\lambda(r) r^{\lambda+2} \mathrm{d}r \right]^2 \tag{10.10}$$

and

$$B(M\lambda) = \frac{\lambda}{\lambda+1} \frac{2J_f + 1}{2J_i + 1} \left[\int_0^\infty J_{\lambda\lambda}(r) r^{\lambda+2} \mathrm{d}r \right]^2 . \tag{10.11}$$

The selection rules for the various terms in (10.1) are such that for natural-parity or "electric" transitions, where $\pi_i \pi_f = (-1)^\lambda$, the magnetic form factor F_λ^M is zero. Similarly for unnatural-parity or "magnetic" transitions, where $\pi_i \pi_f = (-1)^{\lambda-1}$, F_λ^C and F_λ^E are zero. If one scatters from a spin-zero nucleus, only the term with $\lambda = J_f$ in each of the sums of (10.1) remains, and in the second sum either F^E or F^M is zero, depending on the nature of the transition. So only a single multipolarity contributes to the excitation of some level and the cross section is determined by the three densities ρ_λ, $J_{\lambda\lambda-1}$, and $J_{\lambda\lambda+1}$ for electric transitions and by just $J_{\lambda\lambda}$ in the case of a magnetic transition.

The continuity equation connects the currents $J_{\lambda\lambda-1}$ and $J_{\lambda\lambda+1}$ to the transition charge ρ_λ:

$$\sqrt{2\lambda+1} \frac{\omega}{c} \rho_\lambda(r) = \sqrt{\lambda} \left(\frac{\mathrm{d}}{\mathrm{d}r} - \frac{\lambda-1}{r} \right) J_{\lambda\lambda-1}(r)$$
$$- \sqrt{\lambda+1} \left(\frac{\mathrm{d}}{\mathrm{d}r} + \frac{\lambda+2}{r} \right) J_{\lambda\lambda+1}(r). \tag{10.12}$$

Thus, in the case of natural-parity transitions one has only two independent densities and one of the three can be eliminated. It is most convenient to eliminate $J_{\lambda\lambda-1}$ by expressing it in ρ_λ and $J_{\lambda\lambda+1}$.

If one describes the nucleus through the motion of individual nucleons, the total transition density can be expressed as

$$\rho_\lambda(r) = \sqrt{\frac{1}{2J_f + 1}} \sum_{a,b} \rho_\lambda^{ab}(r) S_{ab,\lambda}, \tag{10.13}$$

and similarly for the $J_{\lambda\lambda'}(r)$. This means that the densities are sums over single-particle densities weighted by the spectroscopic amplitudes or density matrix elements

$$S_{ab,\lambda} = \sqrt{\frac{2J_f + 1}{2\lambda + 1}} \langle \psi_f \| [a_a^+ \otimes \tilde{a}_b]_\lambda \| \psi_i \rangle. \tag{10.14}$$

Here, a and b denote single-particle orbitals with quantum numbers nlj, while a^+ and $\tilde{a}$ are particle and hole creation operators. The sum over a and b runs over both protons and neutrons. The values of $S_{ab,\lambda}$ must be provided by the nuclear-structure calculation.

Care should be taken that in calculating the $S_{ab,\lambda}$ the same conventions are used as in the calculation of the single-particle densities. The present programs use the following conventions: Wigner–Eckart theorem without $\sqrt{2J_f + 1}$ factor [see (10.9)], $(ls)j$ coupling, no i^l in the single-particle wave functions, and radial single-particle wave functions positive near the origin. With the nonrelativistic form of the charge and current operators, the single-particle densities in the conventions described above are given by

$$\rho_\lambda^{ab}(r) = C_{ab,\lambda} u_a(r) u_b(r), \tag{10.15}$$

where $u(r)$ is the radial part of the used single-particle wave function.

The transition current densities have two contributions. The first originates as convection of the charge of the nucleons, while the second stems from the magnetization of the nucleons. They are given in units of nuclear magnetons $(e\hbar/2m_p c)$ by

$$\begin{aligned} J_{\lambda\lambda-1}^{ab,C}(r) &= C_{ab,\lambda} \frac{1}{\sqrt{2\lambda+1}} \Big[-\sqrt{\lambda} [u_a' u_b - u_b' u_a] \\ &+ \frac{1}{\sqrt{\lambda}} [l_b(l_b+1) - l_a(l_a+1)] \frac{u_a u_b}{r} \Big] ; \end{aligned} \tag{10.16}$$

$$J_{\lambda\lambda-1}^{ab,M}(r) = \mu C_{ab,\lambda} \frac{1}{\sqrt{2\lambda+1}} \frac{(\chi_b - \chi_a)}{\sqrt{\lambda}} \Big(\frac{d}{dr} + \frac{\lambda+1}{r} \Big) u_a u_b ; \tag{10.17}$$

$$\begin{aligned} J_{\lambda\lambda+1}^{ab,C}(r) &= C_{ab,\lambda} \sqrt{\frac{1}{2\lambda+1}} \\ &\times \Big[\sqrt{\lambda+1} [u_a' u_b - u_b' u_a] + \frac{1}{\sqrt{\lambda+1}} [l_b(l_b+1) - l_a(l_a+1)] \frac{u_a u_b}{r} \Big] ; \end{aligned} \tag{10.18}$$

$$J_{\lambda\lambda+1}^{ab,M}(r) = \mu C_{ab,\lambda} \sqrt{\frac{1}{2\lambda+1}} \frac{(\chi_b - \chi_a)}{\sqrt{\lambda+1}} \Big(\frac{d}{dr} - \frac{\lambda}{r} \Big) u_a u_b ; \tag{10.19}$$

$$J_{\lambda\lambda}^{ab,C}(r) = C_{ab,\lambda} \sqrt{\frac{1}{\lambda(\lambda+1)}} (\chi_a + \chi_b - \lambda)(\chi_a + \chi_b + \lambda + 1) \frac{u_a u_b}{r} ; \tag{10.20}$$

$$J_{\lambda\lambda}^{ab,M}(r) = \mu C_{ab,\lambda}\sqrt{\frac{1}{\lambda(\lambda+1)}}$$

$$\times \left[\frac{\lambda(\lambda+1)}{r} + (\chi_a + \chi_b)\left(\frac{d}{dr}+\frac{1}{r}\right)\right]u_au_b ; \qquad (10.21)$$

with

$$C_{ab,\lambda} = (-1)^{\lambda+j_a-\frac{1}{2}}\sqrt{\frac{(2j_a+1)(2j_b+1)}{4\pi}}(j_a\tfrac{1}{2}j_b-\tfrac{1}{2}|\lambda 0). \qquad (10.22)$$

Here $\chi = (l-j)(2j+1)$, and μ is the magnetic moment of the active nucleon in nuclear magnetons. For completeness the formulae have been given for both $J_{\lambda\lambda-1}$ and $J_{\lambda\lambda+1}$. It is easily checked that the forms given obey the continuity equation as long as no velocity-dependent potentials are present. If such potentials are used, the formulae for the current get more complicated. Except in simple cases (see [10.4,5]) it is not clear what current operator to use.

For convenience (and possible later use in DWBA programs) the densities are expanded in a Fourier–Bessel (FB) series. The densities for r greater than a cut-off radius R_c are assumed to be zero, while for $r < R_c$ they are written as

$$\rho_\lambda(r) = \sum_n A_n q_n^{\lambda-1} j_\lambda(q_n^{\lambda-1}r) ; \qquad (10.23)$$

$$J_{\lambda\lambda+1}(r) = \sqrt{\frac{2\lambda+1}{\lambda+1}}\frac{\omega}{c}\sum_n B_n j_{\lambda+1}(q_n^\lambda r) ; \qquad (10.24)$$

$$J_{\lambda\lambda}(r) = \sum_n C_n j_\lambda(q_n^\lambda r). \qquad (10.25)$$

Here q_n^λ is the nth zero of the spherical Bessel function of order λ. The current density $J_{\lambda\lambda-1}$ is then given by

$$J_{\lambda\lambda-1}(r) = -\sqrt{\frac{2\lambda+1}{\lambda}}\frac{\omega}{c}\left[\sum_n A_n j_{\lambda-1}(q_n^{\lambda-1}r) + \sum_n B_n j_{\lambda-1}(q_n^\lambda r)\right]. \qquad (10.26)$$

The formulae given have to be supplemented for some effects. First, the single-particle densities have to be folded with the nucleon charge or magnetization distribution. This can be done most conveniently in the FB parametrization. If X_n are the FB-coefficients for the point density, the convoluted density is described by coefficients

$$X_n' = X_n F_N(q_n), \qquad (10.27)$$

where the q_ns are the ones defined above and $F_N(q)$ is the appropriate nucleon form factor, i.e. the proton- or neutron-charge form factor in case of the transition charge density or the convection-current term in the transition current density, and the magnetic form factor for the magnetization terms.

If the single-particle wave functions have been calculated by means of the Hartree–Fock (HF) approximation, they are not in the rest frame of the nucleus, for which a correction must be applied [10.6]. This can only be done exactly for harmonic-oscillator wave functions, but as HF wave functions generally are not too different from these, the correction is usually calculated in the harmonic-oscillator approximation. Again, the correction can most easily be included in the FB coefficients, which this time have to be multiplied by

$$\exp\left(\frac{q_n^2 b^2}{4A}\right).$$
(10.28)

Here b is the harmonic-oscillator parameter for the nucleus in question and A is its mass number.

Another possibility is that single-particle wave functions are used as determined from transfer or (e,e'p) reactions. One then has to be aware that the origin of the radial coordinate of the wave function is the center of mass of the $A-1$ core, whereas the electron waves are described with respect to the A system. This difference can be taken into account by modifying the radial scale and magnitude of the single-particle wave function used [10.7].

The model space used in nuclear-structure calculations is often too small, which is reflected in an underprediction of the measured charge densities. On the other hand current densities are often smaller than predicted, which has led to the introduction of effective charges and g-factors for both protons and neutrons. The programs have the option to enter such effective values.

Quantitative calculations of cross sections for detailed comparisons with experimental data, especially for heavy nuclei, have to be done in DWBA, where the distortions of the electron waves due to the nuclear charge are taken into account by solving the Dirac equation. Programs for this are available [10.8]. PWBA calculations, however, already give a good indication, especially when high-energy electrons are used. The main effect of the distortions can be incorporated by comparing the measured cross sections at a certain value of q with PWBA calculations at an effective q, which is given by

$$q_{\text{eff}} = q\left(1 + \frac{3}{2}\frac{\alpha Z \hbar c}{E R_{\text{eq}}}\right).$$
(10.29)

This takes into account that the energy of the electron in the nucleus is larger than its asymptotic value owing to the Coulomb attraction. In this formula R_{eq} is the equivalent radius of a hard sphere; in practice one uses $R_{\text{eq}} = 1.12 A^{1/3}$ fm.

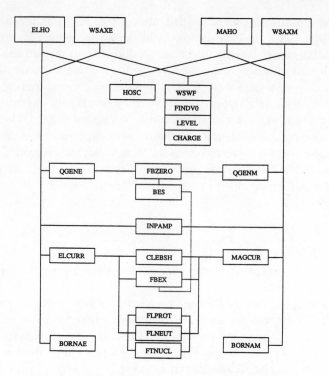

Fig. 10.1. Description of the subroutine structure.

10.3   Description of the Programs

The PWBA programs consist of a set of four programs: ELHO, MAHO, WSAXE, and WSAXM. The label E or M indicates that the program calculates electric or magnetic transitions, while HO or WSAX refers to the use of harmonic-oscillator or Woods–Saxon single-particle wave functions. As the programs overlap in many parts they could have been combined into one program by including a few switches, but for the purpose of clarity this has not been done.

The subroutine structure is as indicated in Fig. 10.1. There follows a brief description of the programs and subroutines.

ELHO, MAHO, WSAXE, WSAXM

The main programs handle all input except the reading of the spectroscopic amplitudes (which is done in INPAMP), call the subroutines HOSC or WSWF to generate the necessary single-particle wave functions, generate the q_n^λ for the FB-expansion, call ELCURR or MAGCUR to calculate the transition densities, which are returned in the form of FB-coefficients, and then calculate in the subroutines BORNAE or BORNAM (in PWBA) the form factors and cross sections. Transfer of variables to subroutines is done partly via parameters and partly via labeled commons.

The center-of-mass correction (according to (10.28) in ELHO and MAHO and following [10.7] in WSAXE and WSAXM) is also calculated in the main programs. The radial step size with which wave functions and densities are calculated has been chosen as 0.1 fm, which is sufficiently small for all practical applications.

HOSC

This subroutine calculates harmonic-oscillator wave functions in steps of 0.1 fm according to the standard polynomial expression; see, e.g., [10.9].

WSWF

The subroutine calculates Woods–Saxon wave functions. If no center-of-mass correction is applied, the grid points are at 0.1 fm intervals. With this correction the wave functions are evaluated at $0.1A/(A-1)$ intervals. The wave functions are solutions of the following bound-state Schrödinger equation:

$$\left(-\frac{\hbar^2}{2m^2}\Delta - V_0 f(r) + V_{\mathrm{so}}\frac{2}{r}\frac{\mathrm{d}f}{\mathrm{d}r}\boldsymbol{l}\cdot\boldsymbol{s} + V_{\mathrm{C}}(r)\right)\phi_{nlj}(\boldsymbol{r}) = E_{nlj}\phi_{nlj}(\boldsymbol{r}).\quad(10.30)$$

The function $f(r)$ is the standard Woods–Saxon shape function

$$f(r) = [1 + \exp((r - R)/a)]^{-1},\quad\quad\quad\quad(10.31)$$

where $R = r(A-1)^{1/3}$. (Notice that A has been defined to be the mass of the target nucleus.) The parameters r and a may be different for the central and spin–orbit potentials. $V_{\mathrm{C}}(r)$ is the Coulomb potential of a homogeneously charged sphere with radius $R = r_{\mathrm{C}}(A-1)^{1/3}$.

Depending on the input the spin–orbit potential is defined as

$$V_{\mathrm{so}} = \left(\frac{\hbar}{m_\pi c}\right)^2 vso\quad\quad\quad\quad(10.32)$$

or, using the Thomas form, as

$$V_{\mathrm{so}} = V_0\left(\frac{\hbar}{2m_{\mathrm{p}}c}\right)^2 vso,\quad\quad\quad\quad(10.33)$$

where vso is the value given in the input.

Again, depending on the input the (negative) energy E is determined for a given V_0 with the help of subroutine LEVEL, or V_0 is determined for given E using the subroutine FINDVO, which uses LEVEL in an iterative procedure. WSWF can also apply a nonlocality correction [10.10] by multiplying ϕ by

$$F_{\mathrm{nonloc}}(r) = \left(1 + \frac{m}{2\hbar}\beta^2 V_0 f(r)\right)^{-1/2}\quad\quad\quad\quad(10.34)$$

and then normalizing it again.

FINDVO

This subroutine determines the value of V_0 for a given value of E. First the energies for an estimated value of V_0 and another value $V_0 + \Delta V_0$ are determined with the help of LEVEL. Then a linear inter(extra)polation is made to obtain the next guess for V_0. This procedure is repeated until convergence.

LEVEL

The subroutine solves the bound-state Schrödinger equation (10.30) for E at given values of the potential parameters. The method used is outward integration from $r = 0$ and inward integration from a value of r well outside the range of the potential and then determination of the Wronskian at some matching radius [10.11].

CHARGE

The subroutine calculates the Coulomb potential of a homogeneously charged sphere.

INPAMP

The subroutine reads the identification of the orbitals a and b involved in a transition and the corresponding spectroscopic amplitude S. The identification can either be given directly via the sequence number of the orbit as used in the actual input or by specifying the charge of the particle and the values of nlj of the orbit. The orbitals given in the actual input are then searched to find the corresponding one. This option is useful when the amplitudes have been generated by some nuclear model. In this case the output of the program that calculates the spectroscopic amplitudes may be tailored to the input of INPAMP.

As mentioned above, one should be careful that the conventions for S in the model calculations are the same as those of the PWBA programs; e.g. spherical harmonics without i^l. As it is not unusual that model calculations include i^l, an option has been included in INPAMP to convert the amplitudes S to those needed in the programs by multiplying them by $i^{l_a - l_b}(-1)$.

Values of S for some simple transitions are as follows: for a pure single-particle transition $\psi_i = |b\rangle$ and $\psi_f = |a\rangle$ one has $S_{ab,\lambda} = 1$; the same value is found if $\psi_i = |0\rangle$ and $\psi_f = [a_a^+ \otimes \tilde{a}_b]_\lambda |0\rangle$, where $|0\rangle$ is full for b and empty for a. The spectroscopic amplitude for elastic scattering is given by $\sqrt{(2J+1)/(2j+1)}\, n$, where J is the spin of the nucleus and n is the number of particles in the orbit with total angular momentum j.

FBZERO

This subroutine returns the nodes q_n^l of the spherical Bessel function $j_l(x)$. They are determined in an iterative process by calculating the value of j_l and its derivative.

ELCURR and MAGCUR

These subroutines calculate the (sum of the) single-particle densities times the spectroscopic amplitudes according to formulae (10.13–17) taking into account possible effective charges and g-factors. If required, the folding with the nucleon form factors may also be performed, following (10.24).

ELCURR does not use the current densities $J_{\lambda\lambda-1}$ and $J_{\lambda\lambda+1}$ internally, but rather the corresponding densities $Y_{\lambda-1}$ and $Y_{\lambda+1}$, which are defined as

$$Y_{\lambda-1} = \sqrt{\frac{\lambda}{2\lambda+1}} \frac{1}{\omega/c} J_{\lambda\lambda-1};$$

$$Y_{\lambda+1} = \sqrt{\frac{\lambda+1}{2\lambda+1}} \frac{1}{\omega/c} J_{\lambda\lambda+1}.$$

It is these Y-currents that are expanded into an FB-series [see (10.24) and (10.26)].

For every single-particle transition the so-called partial $B(E\lambda)$ or $B(M\lambda)$ is also calculated. It is a sort of reduced transition *amplitude* defined as $\int f(r)r^2 dr$, where $f(r)$ is either $\rho_\lambda(r)$ or $J_{\lambda\lambda}(r)$ for this single-particle transition. The values of these (which summed over all single-particle transitions, squared, and multiplied by the appropriate statistical factors yield the $B(E\lambda)$ or $B(M\lambda)$ value) are very useful in finding the relative magnitudes and phases of the various single-particle contributions to a transition. They can also be used advantageously to check the phase conventions of the spectroscopic amplitudes used. Finally the densities are expanded in an FB-series according to formulae (10.23–25).

BORNAE and BORNAM

These subroutines use the FB-coefficients for the charge and/or current densities to calculate these densities, including the $Y_{\lambda-1}$ current in the case of electric transitions, as a function of r in steps of 0.1 fm, according to (10.23–26). Also the $B(E\lambda)$ or $B(M\lambda)$ value is calculated. For $\lambda = 0$ transitions the transition charge and the $M(E0)$ matrix element are calculated, which are defined as $\sqrt{4\pi} \int \rho_0(r)r^2 dr$ and $\sqrt{4\pi} \int \rho_0(r)r^4 dr$, respectively. Then the form factors are calculated according to (10.4–6) for $q = 0.1$ to some qmax in steps of 0.1 fm. For electric transitions F^C and the parts of F^E which are due to $Y_{\lambda-1}$ and $Y_{\lambda+1}$ are given separately. Also for two given angles the energy of the incoming electron that gives the required value of q is determined and for these energy/angle combinations the cross sections are calculated using (10.1) and (10.2). (Note that here the recoil factor is neglected.)

FBEX

The subroutine expands a density $f_l(x)$ into the FB-series

$$f_l(x) = \sum_n X_n j_l(q_n x).$$

The density $f(x)$ must be given in steps of 0.1 fm up to the cut-off radius R_c. The coefficients X_n are returned. They are calculated using the orthogonality relation, which for both cases ($q_n = q_n^l$ or $q_n = q_n^{l-1}$) can be written as

$$\int_0^{R_c} j_l(q_n x) j_l(q_m x) x^2 \mathrm{d}x = \delta_{nm} \frac{1}{2} R_c^3 \left[j_l(q_n R_c)^2 - j_{l+1}(q_n R_c) j_{l-1}(q_n R_c) \right].$$

The integration is carried out by simple trapezoid integration.

BES

This function returns the spherical Bessel function $j_l(x)$. For small arguments the expansion in powers of x is used [10.12]. For large arguments $j_0(x)$ and $j_1(x)$ are calculated in terms of $\sin x$ and $\cos x$. For $l \geq 2$ the function is calculated via the recursion relation

$$j_l(x) = \frac{2l-1}{x} j_{l-1}(x) - j_{l-2}(x).$$

CLEBSH

This function returns the Clebsch–Gordan coefficient $(j_1 m_1 j_2 m_2 | j_3 m_3)$. It is calculated according to Eq. (3.6.11) of [10.13]. The function uses the reduced factorials $n!/a^n$, which are calculated upon the first call of the function. The factor a^n cancels in the calculation of the Clebsch–Gordan coefficients, so that a can be set to any value; its only purpose is to prevent over/underflows in the array of factorials.

FLPROT, FLNEUT and FTNUCL

These functions calculate the longitudinal and transverse form factors of the proton and neutron, which account for the finite size of the charge and magnetization distributions of the nucleons. The transverse form factor is taken to be the same for protons and neutrons. The values for the form factors are taken from [10.14].

10.4   Input Description

The input is from unit "inunit" and the output is written to unit "outunit". The actual values for these are given in a data statement. The spectroscopic amplitudes are either read from "inunit" or from the file "ampfile", where "ampfile" is a character string specified in the input. For later use and plotting purposes the Fourier–Bessel coefficients and the charge and current densities are written to the files: "fbfile", "rhofile", and "curfile".

Unless indicated otherwise, all input is in free format. Some parameters have a default option: if the value entered is zero, the program generates a standard value. For this reason it may be useful to load the programs with a preset value of zero, so that parameters that should get the default value do not have to be entered and an input line such as 0/ may be given. In the following we describe the input of the various main programs.

Input for ELHO

– *title*   (a80)

– *rk, qmax, theta1, theta2*

rk: cutoff radius, default: $1.2A^{1/3}+ 5.0$ fm

note: the FB-coefficients depend on rk.

$qmax$:   max value of q (fm^{-1}) for which cross section is calculated, default: 3.0

theta1/theta2:

angles for which cross section is calculated, default: 90 and 160

– *ifol, dep, den, gpef, gnef, ifolcm*

$ifol$:   > 0: fold with nucleon size

dep/den: additive enhancement of proton/neutron charge, default: 0.0

gpef/gnef: multiplicative enhancement of proton/neutron spin g-factor, default: 1.0

ifolcm: > 0: perform center-of-mass correction

– *norb, amass, azi, bho0*

$norb$:   number of orbits, maximum is 40

if $norb$ < 0 the single-particle wave functions are listed for checking purposes

$amass$:   mass of target nucleus

azi: charge of target nucleus

$bho0$:   harmonic-oscillator parameter for center-of-mass correction, default: $A^{1/6}$

The following lines define the various orbitals. Thus one has to input them *norb* times.

– *chge, an, al, aj, bho*

$chge$:   charge of particle in this orbit

an/al/aj: quantum numbers of this orbit, $an \geq 1$

bho: harmonic-oscillator parameter of this orbit, default: $bho0$

– *inopt*

$inopt$:   input option for spectroscopic amplitudes:

0 = standard, specifying sequence numbers of orbits

1 = standard, specifying quantum numbers of orbits

2 = like 1, but single-particle wave functions defined including i^l

other = other options, to be implemented

– *ampfile*   (a8)

    ampfile: name of input file containing spectroscopic amplitudes;
if blank, continue reading from standard input file

There follows a brief description for the data read from "ampfile".

– *title*   (a80)

    title: only read from external file

– *jinit, parinit, jfin, parfin, lam, etran*

    jinit, jfin: initial and final spin

    parinit, parfin:

        initial and final parity (0 or +1: '+', −1: '−')

    lam: multipolarity of the transition

    etran: excitation energy (zero for elastic scattering is allowed)

If inopt = 0,

– *i, j, ampl*

    i: number of particle orbit

    j: number of hole orbit

    ampl: spectroscopic amplitude for this transition

    repeat until $i \leq 0$ or end of file (EOF) is reached

If inopt = 1 or 2,

– *chgp, np, lp, jp, chgh, nh, lh, jh, ampl*

    chgp/chgh: charge of particle/hole

    np,lp,jp,nh,nl,nj:

        $n, l, 2j$ of particle/hole

    ampl: spectroscopic amplitude for this transition

    repeat until *chgp* < 0 or EOF is reached

Any number of sets can be given starting again with jinit, etc. If EOF
is reached and reading was from "inunit", the program stops. If EOF is
reached and reading was from "ampfile", a new "ampfile" may be given; if
upon reading a new value for "ampfile" EOF is encountered, the program
stops.

Input for MAHO

The input is the same as for ELHO except for line 3, which now reads:

- *ifol, dglp, dgln, gpef, gnef, ifolcm*

> *ifol*:   > 0: fold with nucleon size
>
> *dglp/dgln*: additive enhancement of proton/neutron orbital g-factor, default: 0.0
>
> *gpef/gnef*: multiplicative enhancement of proton/neutron spin g-factor, default: 1.0
>
> *ifolcm*:   > 0: perform center-of-mass correction

Input for WSAXE:

The input is the same as that of ELHO except for lines 4 to 4+*norb*, which now read:

- *norb, amass, azi, r0, a0, vso, rso, aso, rc, beta*

> *norb*:   number of orbits, maximum is 40
>
>   if *norb* < 0, the single-particle wave functions are listed for checking purposes
>
> *amass*:   mass of target nucleus
>
> *azi*:   charge of target nucleus
>
> *r0*:   radius parameter of Woods–Saxon well, default: 1.25

- *norb, amass, azi, r0, a0, vso, rso, aso, rc, beta* (contd.)

> *a0*:   diffuseness of Woods–Saxon well, default: 0.70
>
> *vso*:   < 15: strength of spin–orbit potential, default 6.0
>
>   > 15: Thomas strength of spin–orbit potential, default: 25.0
>
> *rso*:   radius parameter of spin–orbit potential, default: r0
>
> *aso*:   diffuseness of spin–orbit potential, default: a0
>
> *rc*:   radius parameter of Coulomb potential, default: r0
>
> *beta*:   nonlocality parameter, e.g., 0.85

The following lines define the various orbitals. Thus one has to input them *norb* times.

- *chge, an, al, aj, bb, rad*

> *chge*:   charge of particle in this orbit
>
> *an/al/aj*: quantum numbers of this orbit, $an \geq 1$
>
> *bb*:   < 0: binding energy of particle in this orbit, search on potential
>
>   > 0: well depth, search on binding energy
>
> *rad*:   radius parameter for this orbit, default: r0

Input for WSAXM:

The input is the same as for WSAXE except for line 3, which is the same as line 3 of MAHO.

10.5   Output Description

The output is largely self-explanatory. All input parameters are returned in the output, so input errors are readily found. The calculated single-particle wave functions are mentioned and optionally listed. Also the norm and rms radius, calculated by integrating from 0 to rk, are given. The first is useful to check if the chosen value for rk is large enough.

For every single-particle transition the identification is given together with the partial $B(E\lambda)$ or $B(M\lambda)$ value, which as mentioned above is useful for checking the phase conventions of the spectroscopic amplitudes. The total calculated densities are listed, as well as the sum of the partial $B(E\lambda)$ or $B(M\lambda)$ values. Further, the total $B(E\lambda)$ or $B(M\lambda)$ value and the FB-coefficients are given. Finally the form factors and cross sections for the selected values of q are listed.

10.6   Technical Note

The disc contains the main programs ELHO, MAHO, WSAXE and WSAXM as well as a file *ELLIB.FOR* which combines all subroutines necessary to run these main programs. Thus, *ELLIB* has to be linked to all of these programs. Furthermore, the disc contains example input files *ELHO.INP* and *MAHO.INP* for the programs ELHO and MAHO. The files *AMPLI1* and *AMPLI2*, which are examples for external data files "ampfile" containing the spectroscopic amplitudes, must be available in the directory to run the program ELHO if the example input file is used. In the example input file *MAHO.INP* the other option to read in the spectroscopic amplitudes has been used (see Sect. 10.4). In this case the spectroscopic information is included in the standard input file and must not be read from external files. Input files for the other main programs can be generated following the description given in Sect. 10.4. An example output file is found in *OUT.EXA*.

Acknowledgements

The original version of the Woods–Saxon routines is due to Prof. E. Rost (Colorado). One of us (H.P.B.) wishes to thank Dr. J. J. A. Zalmstra for his help in preparing this manuscript.

References

[10.1] J. Heisenberg and H. P. Blok, Annual Review of Nuclear and Particle Science **33** (1983) 234

[10.2] T. W. Donnelly and I.Sick, Rev. Mod. Phys. **56** (1984) 461

[10.3] B. Frois and C. N. Papanicolas, Annual Review of Nuclear and Particle Science **37** (1987) 133

[10.4] J. Heisenberg, Advances in Nuclear Physics **12** (1981) 61

[10.5] L. R. Kouw and H. P. Blok, Phys. Lett. **164B** (1985) 203

[10.6] H. Ueberall, *Electron Scattering from Complex Nuclei* (Academic, New York, 1971) Sect. 3.3.1

[10.7] H. P. Blok and P. D. Kunz, Phys. Lett. **69B** (1977) 261

[10.8] J. Heisenberg, FOUBES1 and FOUBES2 (unpublished programs)

[10.9] P. J. Brussaard and P. W. M. Glaudemans, *Shell-Model Applications in Nuclear Spectroscopy* (North-Holland, Amsterdam, 1977) Eq. (2.21)

[10.10] F. G. Perey and B. Buck, Nucl. Phys. **32** (1962) 353; H. Fiedeldey, Nucl. Phys. **77** (1966) 149

[10.11] S.E. Koonin, *Computational Physics* (Addison-Wesley, Redwood City, 1990)

[10.12] M. Abramowitz and I. A. Stegun, *Handbook of Mathematical Functions* (Dover, New York, 1964)

[10.13] A.R.Edmonds, *Angular Momentum in Quantum Mechanics* (Princeton University Press, 1957)

[10.14] G. G. Simon, Ch. Schmitt, F. Borkowski and V. H. Walther, Nucl. Phys. **A333** (1980) 381

Subject Index

anharmonic vibrator 99, 117
Argonne potential 158, 175
B(E2) values 13, 95, 98, 100, 101, 114, 117pp
B(E3) values 86
B(Eλ) values 192, 199
B(Mλ) values 192, 199
bootstrapping 174
Born approximation 42
Casimir operator 4, 108, 109
center-of-mass correction 29, 34, 48, 197, 201, 203
center-of-mass spurious motion 6
charge density 41pp, 183, 190pp
Clebsch-Gordan coefficients 200
coefficients of fractional parentage 2pp, 89, 92
collective model 81, 85, 88pp, 105pp
collective octupole states 75, 86
collective quadrupole states 75
continuity equation 192, 194
Coulomb energy 32
Coulomb phase shifts 143
Coulomb potential 31, 130, 132, 142, 186, 197, 198, 203
Coulomb sum rule 183
Coulomb wave functions 143, 148
CPU time 167, 181
cranked Nilsson model 51pp
cranking model 52, 58
current density 190pp
Darwin correction 42
Darwin potential 139, 142
detailed balance 174
Dirac equation 130, 138pp, 195
Dirac sea 131
Dirac–Brueckner approximation 138
Dirac–Hartree approximation 129pp
dynamical symmetry 8, 89

effective interaction 1, 7
Efimov effect 152
electromagnetic form factors 171, 190pp
electromagnetic transition rates 13, 20, 22, 95pp, 113pp, 192
electron gas 75
electron scattering 42, 83, 184, 190pp
Euler angles 107
Faddeev equation 152pp
Feynman diagrams 134
Fierz transformation 135, 136
finite difference schemes 28
flux-tube model 171
forced bias 175
form factor
  charge 41, 183, 186, 195
  longitudinal 191, 200
  magnetic 42, 192, 195
  neutron 183, 195, 200
  nuclear 43
  nucleon 134, 195, 199
  proton 183, 195, 200
  Sachs 42
  transverse 191, 200
Fourier components 43
Fourier transform 136, 137, 183, 191
Fourier–Bessel coefficients 195, 200, 201, 204
Fourier–Bessel series 194
Fourier–Bessel transform 42, 44, 191
gamma-unstable nuclei 89, 101
Gauss elimination 40
Gauss–Laguerre quadrature 141
Gauss–Legendre quadrature 158, 163, 169
giant dipole resonance 86
giant quadrupole resonance 85
gradient iteration 15, 38, 39